AQUACULTURE AND
WATER RESOURCE MANAGEMENT

AQUACULTURE AND WATER RESOURCE MANAGEMENT

EDITED BY

DONALD J. BAIRD
MALCOLM C.M. BEVERIDGE
LIAM A. KELLY

and

JAMES F. MUIR

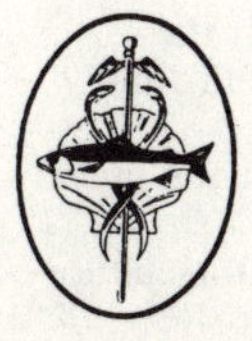

INSTITUTE OF AQUACULTURE

Blackwell Science

Contents

Preface

The increasing global importance of aquaculture as a food-producing activity is mirrored by the continuing decline in wild fish stocks. With this explosion in aquaculture production have come attendant environmental problems. Owing to its tight linkage with natural ecosystems, it can be argued that aquaculture has the potential to have a much more profound impact on environmental quality than terrestrial farming systems of similar size. Moreover, it can also be argued that to become sustainable, practitioners of aquaculture must place more emphasis on recognising these linkages, and protecting the ecosystems which nurture and sustain their industry.

To address these issues, a conference, titled 'Aquaculture and Water Resource Management' was held at the University of Stirling from 21st to 25th June 1994, and was attended by over 120 delegates from more than 20 different countries. Many of the contributed papers presented at the conference have now been published (*Aquaculture Research*, Vol. 26, issues 8 and 9, 1995). This present volume contains the papers presented by eight 'scene-setting' invited speakers, dealing with a range of scientific and management issues relating to the conference topic.

Like the conference, the aim of this book is to stimulate debate within the aquaculture and water management community, including researchers, managers and operators, to provide evaluation of the current problems posed by the industry. The book also considers real and perceived environmental problems associated with aquaculture within a wider environmental context, including the means to remove or otherwise mitigate their impact. The contributing authors are all recognised experts in their specific fields.

The first chapter by Robin Welcomme sets the scene, by considering aquaculture as a global activity which is facing significant increases in consumer demand due to dwindling natural stocks. James Muir offers a systems and sustainability perspective for aquaculture and the environment in Chapter 2. In Chapter 3, Asbjørn Bergheim and Torbjørn Åsgård deal with the problems arising from waste output from intensive aquaculture, illustrating one of the environmental costs associated with aquaculture, the generation of waste which inevitably results from intensive livestock production. Chapter 4 by Barry Costa-Pierce and Chapter 5 by Angela Arthington and David Blühdorn illustrate clearly that the interactions of aquaculture in the environment extend beyond the impact of wastes on aquatic ecosystems: the most far-reaching effects may be those which result through species interactions, either between native species cultured in large numbers, or as a result of exotic introductions released either inadvertently or deliberately into the wild. In Chapter 6, Donald Weston critically examines the poorly-studied and controversial area of release of animal health treatments, including pharmaceuticals, into the environment around aquaculture systems, currently a major concern amongst both producers and opponents of aquaculture.

The final two chapters shift the focus away from the nature and extent of aquaculture impacts towards a consideration of how these impacts can be controlled, both in terms of on-farm practices and through legislation. In Chapter 7, Simon Cripps and Liam Kelly critically review the technology available for reduction of wastes from intensive aquaculture, and how these may be employed more effectively in the future. Finally, in Chapter 8, Bill Edeson discusses the degree to which effective control can be implemented by external regulation, with reference to aquaculture legislation in different parts of the world.

In organising a conference of this type, the efforts of many people are required to identify appropriate topics for discussion and relevant speakers. We extend warm thanks to those involved in the scientific organisation of the meeting, Richard Gowen, David Mackay, Donald McClusky and John Stellwagen, and thank John Roberts and Trouw Aquaculture for their sponsorship of the meeting.

Donald J. Baird, Malcolm C.M. Beveridge, Liam A. Kelly and James F. Muir
Institute of Aquaculture
University of Stirling
Stirling
UK

Chapter 1
Aquaculture and World Aquatic Resources

Robin L. Welcomme, *FAO, Rome, Italy*

1.1 INTRODUCTION

Current rates of increase in world population are calling into question the capacity of humanity to continue to ensure a supply of food for all into the next century. Human population increase is such that estimates for the year 2000 are 6 261 000 000, of which 80% are in the developing countries. In 2025 these figures will have risen to 8 500 000 000 and 84% respectively (UN 1991). At the same time agricultural production has stagnated over the last 2 years after a period of impressive growth through the 1970s and 1980s (FAO 1993). The issue of food security has now become crucial in planning for future generations, and the fisheries sector, along with other agricultural activities, is being examined to determine the sustainable contribution it can make to future food supply. This paper examines current cropping patterns on the worlds aquatic resources and attempts to identify future trends of the capture and aquaculture sectors.

1.2 WORLD TRENDS IN FISH CATCH

1.2.1 General situation

The fisheries sector appears to be following the same pattern of stagnation as most other food-producing activities. The nominal catch statistics of FAO (FAO 1993) indicate that overall production has stabilised and even declined over the last 3 years following a long period of almost uninterrupted growth (Fig. 1.1). Preliminary figures for 1992 indicate an increase of 1 100 000 t over 1991, but this is still below the peak of 1989.

Fish is one of the major sources of animal protein world wide. In 1990 it accounted for 16% of supply globally having risen from 14% in 1961. The percentage contribution is somewhat lower in developed countries (13.8%) than in developing countries (19%) (Laureti 1992). The contribution of fish to global per caput food supply rose until 1989 when it began to fall (Fig. 1.2). A considerable proportion of captured marine fish is converted to fish meal and does not enter directly into human

1

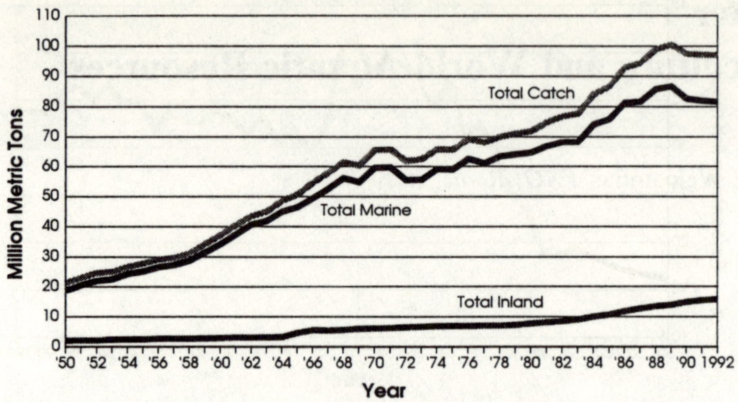

Fig. 1.1   World total catch of fish, crustaceans and molluscs, 1950–1992.

consumption. Rather it is utilised as a component of animal feed for poultry, pigs, fish and shrimp and sometimes as fertiliser. Before the collapse of the Peruvian anchovy in 1970 the proportion used in this way reached 40% but has since stabilised at about 30% (Fig. 1.3). It is still too early to determine whether the falls in production and per caput supply represent a long term trend or are a short term manifestation of social and political changes. There are strong indications that capture fisheries have reached a level of stabilisation and that substantial increases over and above present levels are unlikely.

To compensate for the lack of potential for substantial increase in marine fisheries there is a tendency to try to locate alternative stocks, particularly in the deeper parts of the sea where past estimates of meso-pelagic abundance are particularly high. Despite the apparent abundance of stocks of this type, they have not generally shown themselves suitable for exploitation with existing technologies. Owing to their small

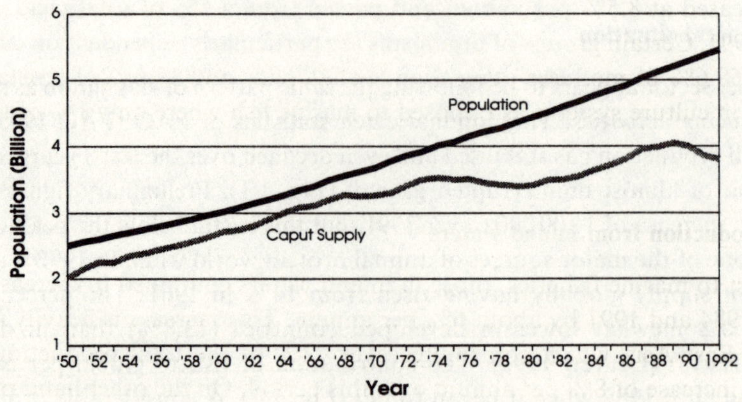

Fig. 1.2   Global per caput food fish supply, 1950–1992.

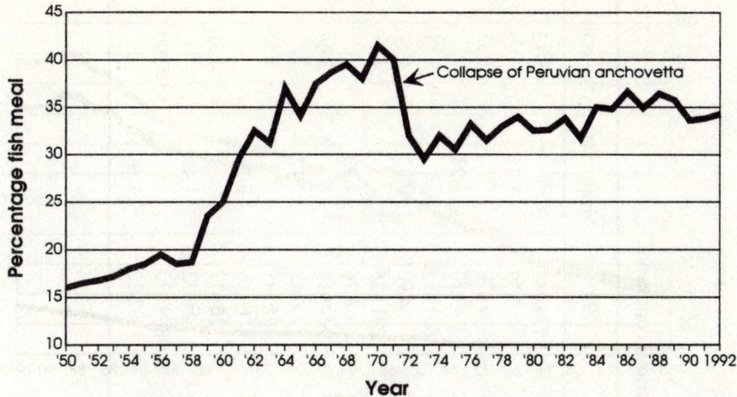

Fig. 1.3 Percentage of nominal marine fish catch converted to fish meal, 1950–1992.

size and other factors meso-pelagic species are generally unsuitable for direct human consumption and would have to be converted into other non-food products such as fish meal. As a result, their exploitation is probably not cost effective at present, although with estimates of stock size running into hundreds of millions of tonnes there is undoubtedly some potential to increase overall production of marine fish if only for use as animal feed.

1.2.2 Production from marine waters

Examination of the trends in Fig. 1.1 shows the origin of yields from marine environments. In 1991, the majority of this production (94.7%) arose from marine capture fisheries and the balance from aquaculture. The trends in the two sectors are different, however. Since 1985 capture fisheries increased by 1% per annum, declined in 1990–1991 and increased again very slightly in 1992. Mariculture production on the other hand increased at 8.5% per annum and passed from 3.5% of total yield in 1985 to 5.3% in 1991. Certain groups of organisms are particularly dependent on culture. For example, 58.6% of molluscs other than cephalopods and 13.6% of crustaceans are produced in culture systems as opposed to marine fish where only 1% of the total is cultured.

1.2.3 Production from inland waters

In contrast to marine fisheries, those in inland waters continued to increase globally between 1984 and 1991 by about 6% per annum. This increase is heavily influenced by South East Asia which contributes over 72% of the total production and has shown an increase of 8% per annum over this period. On the other hand production from Europe and South America had similar yields in 1984 and 1991 (Table 1.1).

Table 1.1 Production, percentage yearly change and percentage contribution of inland capture fisheries in the various continents.

	1984	1985	1986	1987	1988	1989	1990	1991	Mean rate	% total
Asia	2 949 635	2 835 900	2 993 959	2 931 770	2 958 193	3 229 439	3 525 845	3 786 444		55.12
% change		−3.86	5.57	−2.08	0.90	9.17	9.18	7.39	3.75	
Africa	1 452 504	1 443 674	1 580 510	1 639 367	1 729 989	1 729 905	1 867 624	1 809 670		26.35
% change		−0.61	9.48	3.72	5.53	0.00	7.96	−3.10	3.28	
Former USSR	614 734	616 297	607 417	643 454	636 257	669 017	576 625	604 891		8.81
% change		0.25	−1.44	5.93	−1.12	5.15	−13.81	4.90	−0.02	
America, S	325 772	315 878	348 446	368 964	335 151	314 712	303 231	291 825		4.25
% change		−3.04	10.31	5.89	−9.16	−6.10	−3.65	−3.76	−1.36	
America, N	225 749	225 070	226 282	294 996	270 474	255 051	266 369	249 918		3.64
% change		−0.30	0.54	30.37	−8.31	−5.70	4.44	−6.18	2.12	
Europe	172 162	156 576	172 423	140 362	153 897	138 268	125 348	105 345		1.53
% change		−9.05	10.12	−18.59	9.64	−10.16	−9.34	−15.96	−6.19	
Oceania	19 110	19 555	19 867	21 150	21 661	21 910	22 392	20 977		0.31
% change		2.33	1.60	6.46	2.42	1.15	2.20	−6.32	1.40	
Total	5 759 666	5 612 950	5 948 904	6 040 063	6 105 622	6 358 302	6 687 434	6 869 070		100.00
% change		−2.55	5.99	1.53	1.09	4.14	5.18	2.72	2.58	

Changes within the sector demonstrated the growing importance of aquaculture, which contributed 42% of the total in 1984 and 55% in 1991. Aquaculture rose globally by over 6% per annum and the slowest growth registered was that of Europe at 3.25% per annum (Table 1.2). In contrast capture fisheries remained static or declined on several continents. In the waters of the former USSR catches remained unchanged, in South America they declined by about 1.36% per annum and in Europe losses in catch amounted to 6.2% per annum. Other continents recorded increases, with Asia (3.75% per annum) and Africa (3.28% per annum) as the leaders. The overall rate of increase may be slowing, as all continents except Asia registered decreases in 1991. In Asia the figures are distorted by China, which, with a 1991 catch of 1 232 534 t, contributes 18% of global and 33% of Asian catches. Of 33 Asian countries, seven remained more or less stable from 1984 to 1991 and nine registered declines.

1.3   CAPTURE FISHERIES

The stabilisation and decline of capture fisheries throughout most of the world can be traced to three possible causes, fishing pressure, environmental degradation and socio-economic factors, which may operate in isolation or together.

1.3.1   Fishing pressure

There is clear evidence that most major stocks of marine fish are either fully fished or overfished. Furthermore, until recently there has been a continuous replacement in the fishery of high value, large, traditional species by lower value, shorter-lived ones (FAO 1992). The capacity of marine stocks to continue to support this fishing-down process is clearly limited. The accessibility of inland waters to large numbers of people has facilitated the exploitation of most stocks at levels approaching or exceeding the optimum. Thus the phenomenon of fishing down the usually rich and varied fish communities of inland waters is advanced in all regions. A typical example of this process is that of the Oueme river in Benin, where heavy fishing pressure has led to the elimination of several species and has transformed the community from one based on large, long lived species to one whose principal components are all small and short-lived (Fig. 1.4). That fishing pressure is responsible is in little doubt as the Oueme retains its pristine riverine ecology. Despite this example rivers continue to sustain high levels of off-take owing to the extraordinary resilience of the fish communities to climatic variability, which serves as a pre-adaptive mechanism to heavy fishing pressure.

1.3.2   Environmental degradation

Until recently there has been an emphasis on pollution as the major factor endangering aquatic organisms. The process of eutrophication also attracted a good deal of

Table 1.2 Production, percentage yearly change and percentage contribution of aquaculture in the various continents.

	1984	1985	1986	1987	1988	1989	1990	1991	Mean rate	% total
Asia	3 464 481	4 196 650	4 859 838	5 647 965	6 252 711	6 400 776	6 834 631	7 162 282		86.21
% change		21.13	15.80	16.22	10.71	2.37	6.78	4.79	11.11	
Former USSR	266 747	289 345	319 510	344 927	359 330	350 673	398 289	425 890		5.13
% change		8.47	10.43	7.95	4.18	−2.41	13.58	6.93	7.02	
Europe	270 401	274 981	286 987	308 387	324 502	339 179	344 620	337 212		4.06
% change		1.69	4.37	7.46	5.23	4.52	1.60	−2.15	3.25	
America, N	208 414	218 857	258 446	278 282	261 865	282 400	272 612	283 594		3.41
% change		5.01	18.09	7.68	−5.90	7.84	−3.47	4.03	4.75	
Africa	23 786	38 883	39 310	42 697	54 525	73 984	51 361	58 255		0.70
% change		63.47	1.10	8.62	27.70	35.69	−30.58	13.42	17.06	
America, S	12 716	13 118	13 674	17 454	21 481	25 288	31 582	38 644		0.47
% change		3.16	4.24	27.64	23.07	17.72	24.89	22.36	17.58	
Oceania	1 070	1 145	980	1 685	2 041	1 865	1 960	2 242		0.03
% change		7.01	−14.41	71.94	21.13	−8.62	5.09	14.39	13.79	
Total	4 247 615	5 032 979	5 778 745	6 641 397	7 276 455	7 474 165	7 935 055	8 308 119		100.00
% change		18.49	14.82	14.93	9.56	2.72	6.17	4.70	10.20	

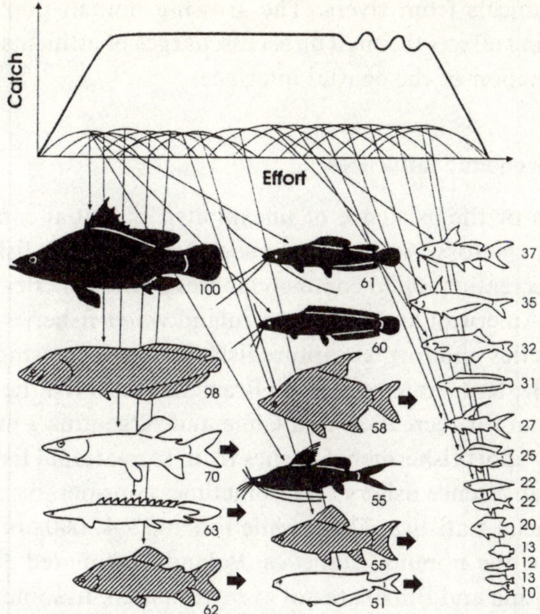

Fig. 1.4 Succession of species in the Oueme River fishery, 1957–1971. Individual species illustrated usually represent blocks of similar species (e.g. some six Synodontis are known from the Oueme although only one is illustrated). Species shown in black had disappeared completely from the fishery by 1965. Those shown hatched were present in reduced numbers. Thick arrows indicate species present at much smaller size. Figures are the total lengths of the species or groups.

attention and it became clear that the effects of nutrient enrichment in many ways paralleled that of the fishing-down process. A third factor, structural alterations to the form and function of the environment, has increasingly been implicated in the modification of fish communities. Because of the similar and probably synergistic relationships between these various factors, confusion often arises as to the true source of decline in fished communities. By and large, both fishing pressure and environmental changes can be implicated except in those few systems where relatively pristine ecological conditions exist or where fisheries only impact in an incidental way on the aquatic communities.

The decline of fish communities in inland waters due to pollution was particularly marked during the phase of industrialisation in the temperate zone and resulted in the disappearance of many species. This tendency has continued into newly industrialising nations, and many of the rivers in countries such as Brazil, China and India have faunas depleted by a combination of falling water quality and environmental modification.

Coastal and enclosed marine environments such as the Adriatic and the Black Sea are also beginning to show signs of moderate to severe degradation (Caddy 1993) (Fig. 1.2), much of which can be traced to land-based discharges of toxic and eutrophicating chemicals from rivers. The growing human populations in coastal areas are also exerting effects through direct discharges of effluents and modifications in the form and function of the coastal interface.

1.3.3   Social and economic influences

Falls in production or the existence of unexploited potential can also be traced to social and economic factors. An important social influence on fisheries originates in conflicts between recreational and commercial uses of the fisheries resource. Many of Australian, North American and European inland water fisheries are today devoted to recreational fisheries and formal capture fisheries are increasingly rare. This trend is not confined to the temperate regions as Brazil has reserved the 100 000 t potential of the Gran Pantanal for recreation and Chile and Argentina's inland resources are mainly exploited by sport fishermen. Catches from recreational fisheries often form a type of concealed subsistence fishery. The sometimes considerable tonnages are rarely recorded in the official statistics. Thus, while Finland's 47 000 t of recreational catch (1992) do appear in the nominal statistics, Poland's estimated 38 000 t do not and catches from Argentina and Brazil are not even estimated. In some areas of the world, sectors of public opinion are seeking to ban even recreational fishing, using conservational and animal rights arguments.

1.4   AQUACULTURE

It has often been suggested that aquaculture will expand to compensate for shortfalls from catches. Certainly, the present trend, with aquaculture as the only fisheries sector to be increasing, would appear to support this contention. The continuing rise in production from aquaculture is paralleled by a diversification of the sector, with more and more species entering culture systems. For example, FAO (Food and Agriculture Organisation) statistics list an increasing number of statistical categories from 1984 to 1992 (Table 1.3). However, this diversification is taking place within a fixed resource base and, as yet, no new form of culture has been introduced which would allow for a substantial jump in potential.

The period of rapid expansion of most forms of aquaculture of the early 1980s seems to be over and culture for the current species appears to be slowing. The reasons for this seem to be that aquaculture has contemporaneously hit a series of constraints, some of which are inherent in the intensification process and others which are external to the sector.

1.4.1   Limits imposed by disease

Like all farmed livestock, fish, molluscs and crustaceans raised under semi-intensive and intensive farming systems are highly prone to stress due to proximity, handling

Table 1.3 Number of categories of organisms listed in FAO nominal statistics.

a) used for inland aquaculture

	1984	1985	1986	1987	1988	1989	1990	1991
Crustaceans	14	15	15	14	16	16	17	17
Fish	72	73	81	80	83	84	91	96
TOTAL	86	88	96	94	99	100	108	113

b) used for marine aquaculture

	1984	1985	1986	1987	1988	1989	1990	1991
Crustaceans	14	18	17	20	21	19	19	20
Fish	37	39	46	49	49	47	53	55
Molluscs	40	41	41	46	48	49	51	51
Others	3	3	3	3	3	3	3	3
TOTAL	94	101	107	118	121	118	126	129

and inappropriate water management. They are consequently very susceptible to outbreak of disease and mass mortalities. Not surprisingly the trend to increasing disease is particularly marked in south-east Asia. For example, in 1993, Epizootic Ulcerative Syndrome (EUS) resulted in economic losses of over $US15 million in Bangladesh, Nepal, Sri Lanka and Thailand alone. This is the latest chapter in a ten year history of the disease which has plagued cultured and wild stocks in south-east Asia and has penetrated as far as China and India. What is particularly notable about EUS is that it is one of a number of such diseases afflicting the aquaculture sector for which a remedy is not known. Also in 1993, over 146 000 ha of shrimp farms in China were struck by various unidentified viruses causing an economic loss of $US172 million (figures from FAO-GLOBEFISH database). Problems of this type are not limited to the south-east Asian region as viral infections of shrimp have caused problems in the Gulf of California, where many farms had to close as a consequence. Closer to home a parasite introduced into Norway with Atlantic salmon from the Baltic now threatens wild Norwegian stocks of the species and the Neva salmon of the Baltic is similarly afflicted. It can be expected that in the absence of measures to improve husbandry and without stricter controls on water quality and movement of species, the incidence of disease outbreaks will increase with further intensification of culture.

1.4.2 Limits imposed by intensification of land use

The growing world population is already exacting a toll on the availability of land for agriculture. Much of the current stagnation in agricultural production has been traced to population pressure and land degradation, which are undermining agriculture's current condition and future prospects (World Resources Institute 1992). For instance, about 10% of world vegetated lands are listed as having been degraded since 1945. This situation is locally severe in Central America (25%), Europe (17%), Africa (15%) and Asia (12%). To aggravate this situation most of the areas of arrested growth and even decline in production are in the developing countries where population growth is maximal. In areas of high population growth the division of existing land into smaller and smaller plots limits the area available to the individual farmer to adopt alternative cropping strategies such as aquaculture. Similar stresses are appearing in coastal zones where population growth is particularly high and where much of the land suitable for development of shore-based installations has already been cleared and developed for various types of culture. The uncontrolled spread of shrimp culture in many former mangrove areas has already resulted in land degradation and the abandonment of formerly productive sites.

The regions suitable for aquaculture are perhaps more limited than previously supposed and studies such as that for Africa (Kapetsky, 1994) indicate that the combination of soil type, water availability, proximity to markets and sources of feedstuffs, etc., are restricted to certain well defined areas. Clearly, even within such suitable territory, aquaculture is going to have to compete with the many other uses to which land, water and nutrient resources can be put, and as population pressures increase it is to be expected that priority will be given to cereal crops and vegetable protein sources. Particular competition exists between irrigation and aquaculture for use of water. In general, the idea of combining some form of aquaculture within irrigated areas, through cage culture in canals or through the inclusion of fish ponds in the irrigated perimeter, seems appealing, but little success has been achieved in attempts to mix the two technologies. Indeed, irrigation successively salinises the areas inundated, especially in arid regions, and eventually causes losses of both land and water.

1.4.3 Limits imposed by water demand

Availability of clean water is fast becoming one of the principal limiting factors for life. As indicated in Table 1.4, by the year 2000 water availability per caput will have fallen to 25% of its 1950 value in Latin America and Africa, to 30% in Asia and 50% in North America. Only in Europe has supply remained relatively constant due to low population growth. The need to conserve water quality works both ways, for not only is clean water needed for aquaculture, but also the effluents from intensive aquaculture installations are themselves a source of pollution. This has two principal consequences. Firstly, the numbers of installations on a given stream or water body have to be limited and many of the guidelines that are now being formulated are aimed at defining such limits. Secondly, the need to conform to water quality regulation imposes costs on the aquaculturist that may prove excessive in economically marginal operations.

Table 1.4 Per caput water availability by region, 1950–2000.

Region	1950	1960	1970	1980	1990	2000
Latin America	105	80.2	61.7	48.8		28.3
North America	37.2	30.2	25.2	21.3		17.5
Africa	20.6	16.5	12.7	9.4		5.1
Europe	5.9	5.4	4.9	4.4		4.1
Asia	9.6	7.9	6.1	5.1		3.3

1.4.4 Limits imposed by availability of feedstuffs

The different types of aquaculture can be considered as falling into two broad categories irrespective of the social infrastructure supporting it: culture of high value, carnivorous fish and shrimp species based mainly on supplies of fish meal and trash fish as feed inputs, culture of lower food chain, and culture of usually lower value species based on locally available vegetable wastes. As we have seen, some 30% of marine fish catch is converted into fish meal. The resource base for this product is unstable and has already suffered one major collapse in 1970. Stocks of pelagic fishes upon which the fish meal industry is based are already fully exploited world-wide, and it is evident that supplies of fish meal are limited. The high-value sector that depends on fish meal is thus equally finite save for the unlikely eventuality that some other form of cost-effective high protein feed can be located. Feeds for the lower food chain species, on the contrary, are more abundant, and some can even utilise enhanced natural productivity in extensive culture systems. It follows then that the best potential for future advances in culture should be obtained using lower food chain species such as carp and tilapia. Current expansions in the rearing of these species are however limited by consumer preferences and other social and economic factors in some regions of the world.

1.4.5 Social and economic constraints

Experience has shown that social and economic constraints are among the most serious limitations to the goal of diversifying rural agricultural options and diets by introducing aquaculture into areas where potential is good but practice is so far lacking. Whereas technical solutions are usually available even under marginal conditions, problems such as overcoming traditional land tenure patterns, finding time for a new activity in already busy schedules, influencing perceptions of risk and food security or allocating scarce financial resources have proved far less tractable. Where options open to rural communities are squeezed by increasing pressure on land, water and other commodities through population growth, the intractability is likely to harden rather than disappear.

Changes in consumer patterns are influencing the nature and the distribution of the products of fisheries. Areas where fish are not accepted as a regular item of diet were relatively common, but with increasing education and shortages in food supply such cases are becoming rare. At the same time, the trend to urbanisation is creating demand for more varied diets, including fish, among city dwellers. This involves a transfer of fishery and aquaculture products from rural to urban sectors to the detriment of the rural sector. Furthermore, the increase in demand by the richer nations for health reasons, coupled with their higher purchasing power, means that much fish caught in the poorer regions of the world is exported in the form of value added products.

Recent developments of aquaculture have shown the sector to be especially sensitive to such influences. The most widely publicised successes in recent years have been precisely in those high value sectors such as shrimp, salmon and marine carnivores which most appeal to urban markets of industrialised countries and which have attracted the bulk of research funds. The publicity and the volume of scientific output through publications and meetings, however, frequently obscure the fact that 80% of world aquaculture of fish and crustaceans is for lower value, low food chain species, principally carp.

Unfortunately carp are not to everyone's taste and culture systems adapted to the conditions peculiar to China are not readily transferable elsewhere. Even in eastern Europe, where carp is a traditional dietary item, there is a trend to forsake it for more attractive salmonids where the opportunity is granted. Tilapia, however, are much more acceptable and could well form the basis for expansion of fish culture in Africa and South America. The failure of such an obviously attractive means of producing cheap protein to establish itself in either continent derives from a complex of social and economic influences and land-use patterns. Expansion of aquaculture in western and southern Africa is currently hampered by competition with cheap fish of marine origin, and even were cheap fish to be available in South America consumer preference is still strongly oriented to red meat. Should these influences change, production levels consistent with potential might be attained in both these areas.

1.5   INTENSIFICATION

With the apparent saturation of the capacity to support exploitation by inland and marine waters, there has been an increasing trend to search for and apply measures to increase the productivity of natural and semi-natural systems. Two main approaches to intensification have been adopted, stock enhancement and ecosystem enhancement. The two approaches are linked in that the intensification process is very variable. Initially stocking and enhancement techniques are used to support a recreational or capture fishery. With further intensification, stocked waters are managed as extensive culture systems. The final stage comes with the introduction of intensive culture systems such as cages or isolated coves managed through combinations of stocking, ecosystem enhancement and fertilisation, which may co-exist with more traditional capture activities.

1.5.1   Stock enhancement

Figures for stocking into marine and freshwater environments are reported to FAO by the countries of the world. Unfortunately the data are not consistent from year to year and cannot be used to establish trends or the relative predominance of the practice. For instance, neither China nor India, countries heavily dependent on stocking for the management of their fisheries, have included data on stocking in their reports. Some numbers suffice to illustrate the scale of stocking in some areas of the world. Europe, for instance, stocks over 2 billion (2×10^9) fry per year, the ex-USSR 8 billion (8×10^9), with numbers as high as 30 billion (30×10^9) in some years, and Bangladesh 6 billion (6×10^9). The poor nature of the figures provided to the FAO data base, however, means that evaluations have to be derived from more general literature.

Enhancement of inland stocks

Stock enhancement of inland waters has been practised at least since the end of the nineteenth century and in some areas long before. Stocking is used as a management tool mainly to maintain species that do not breed naturally in the recipient water, and to support species that can breed naturally but are fished at levels beyond the natural reproduction rate or in which environmental degradation has limited the capacity for recruitment.

Many European and North American systems have fisheries which have been sustained for several decades by stocking species of salmonid, coregonid and cyprinid (see several papers in van Densen *et al.* 1990; Moehl & Davies 1993). In the tropics, intensification of the productivity of fisheries in this manner is especially common. In Asia, stocking is an established practice but is rarely quantified (Baluyut 1983) although in Sri Lanka some 9 000 000 fry were stocked into reservoirs in the period 1980–1988 (De Silva 1988). In China, particularly, the regular stocking of dams and reservoirs to levels equivalent to extensive aquaculture is a cornerstone of management policy (Li 1988; De Silva *et al.* 1992). More recently Chinese reservoirs have been managed even more intensively through combinations of culture-based and culture fisheries (Lu 1992). Further increases in productivity have been induced in smaller water bodies by fertilization or combinations of fish–duck–pig rearing. Such systems have increased productivity about tenfold and now run as high as 1800 kg ha^{-1} in the main water body and 500 kg ha^{-1} in coves. This trend to intensification of yield may account for the sustained increase in inland fish production noted from China.

Intensively stocked systems are still comparatively rare in Africa, although the potential for expanding fish production in small water bodies is being explored both in west and southern Africa through a series of projects. Burkina Faso, for example, has a permanent natural water resource that has been badly affected by the Sahelian drought. Nevertheless the country has been able to expand its annual yield from about 5000 t per annum in 1975 to about 8000 t per annum in 1990, largely by the exploitation of more than 300 dams created for agricultural purposes. Management is still unsympathetic, but even this limited success indicates that areas such as the Sahel, rich as they are in small artificial water bodies, can produce fish for local needs.

Systematic stocking of small lakes and reservoirs with local species and introduced tilapia is widespread in Latin America and is particularly well developed in Cuba, Mexico and parts of Brazil (Juarez-Palacios & Olmos-Tomassini 1992; Juarez Palacios & Varsi 1993). The figures from Cuba (Fig. 1.5) show clearly the relationship between the intensification of stocking and the increases in yield which reached 22 000 t in 1992 based entirely on small dam fisheries (Mari-Diaz 1989). Most of the production is tilapia, although in recent years Chinese carp have been introduced to diversify the fishery. Similar series of figures have been produced for reservoirs in Mexico where mobilisation of the many dams in the country through stocking has raised production steadily from virtually zero to 168 000 t in 1992. In some cases Mexican reservoir productivity is based on a single introduction of a new species, often a tilapia, whereas in others continuous stocking is needed. In the drought polygon area of north-east Brazil, some 5.5 million fingerlings of local and introduced species were stocked per year over 9 years from 1980 to 1988 into an average of 1500 dams to produce about 17 000 t of fish per year. Of this production, 78% was due to introduced species, of which 30% were tilapia (Juarez-Palacios & Olmos-Tomassini 1992).

The practice of stocking into rivers where the major species are migratory and where the normal pathways are interrupted by dams has been relatively well established for many years. Commercial fisheries are frequently supported in this way. For example, 90 000 000 young sturgeon and 18 million *Stenodus leucichthys* are stocked annually into the Volga river (Pavlov & Vilenkin 1989) and without such support

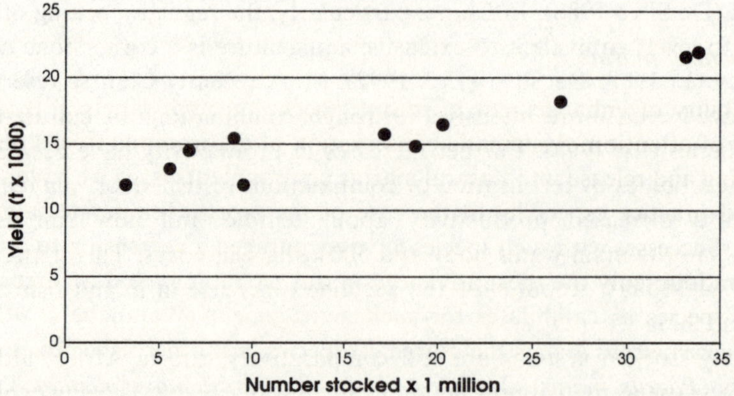

Fig. 1.5 Correlation between numbers of fry stocked and yield at fishing 2 years later in a series of reservoirs in Cuba.

these species would doubtless disappear. Most salmon fisheries of the western North American rivers are similarly supported. The Columbia river, for instance, was stocked with 395 000 000 young salmonids in 1983 (Ebel *et al.* 1989).

One of the major aspects of stocking as a management tool that needs investigation is the carrying capacity of aquatic ecosystems in the face of massive injections of recruits. Despite the world-wide popularity of sometimes massive stocking programmes, few systematic evaluations have been made either of the effectiveness of these practices in raising the production/productivity of the system and their cost–benefit ratios, or of their impacts on local fish communities. For instance an application of population dynamic theory to enhanced fish communities is only just emerging (Lorenzen 1995).

Similar effects to constant stocking programmes can in some cases be achieved by simple introductions. Transfers of species outside their natural range have been widespread as an enhancement tool, particularly in the former USSR where a number of food organisms such as mysids were transferred together with fish to improve productivity of water bodies with poor indigenous stocks. Similarly, many of the gains in productivity in Brazilian and Mexican lakes and reservoirs followed the introduction of exotics such as tilapia, carp or black bass without the need for further interventions. Such introductions are often associated with boom and bust cycles in which populations of the new species expand to extremely high levels only to subside after some years. For example, in Sulawesi successive waves of introductions of species such as *Puntius gonionotus*, *Oseochilus hasselti*, *Helostoma temminki* and *Oreochromis niloticus* raised the yield from the Lake Tempe complex from its original 2000 t per annum to as high as 48 000 t in the 1940s. Productivity has subsequently dropped to about 15 000 t per annum, which seems to correspond to the sustainable level of productivity (Petr 1994; Petr in press). Similar histories have been noted in Thailand where populations of *Oreochromis niloticus* in reservoirs form the basis for large fisheries, which last for about 5 years before they decline.

Enhancement of marine stocks

The possibility of enhancement of marine stocks through ranching is attracting a good deal of attention. Systematic conservation of diadromous stocks through artificial rearing and release into natural marine environments is, as we have seen, a long established practice especially in the case of the North Temperate salmonids and sturgeons. Successes with such species have encouraged an extension to other species. Japan is undoubtedly the most advanced in this technology and is currently investigating 57 species as candidates for such management (Watanabe *et al.* 1989). At present *Oncorhynchus keta* is the dominant species, but other fish species such as red sea bream (*Pagrus major*), Japanese flounder (*Paralichthys olivaceus*), black sea bream (*Acanthropagrus schlegeli*) and rock fish (*Sebastes schlegeli*) all involve releases of over 1 million fry. Efforts are not limited to fish as stocks of molluscs and crustaceans such as scallops, abalones, crabs and Kuruma prawns are enhanced by extensive stocking. Salmonid management programmes are also found in Russia, USA, Canada, Iceland, Norway and Finland (Isaksson 1988). An increasing number of experimental releases of cod species are being made in the USA, Canada and Norway and a range of other fish species, red drum, turbot, sole and plaice, are being investigated as candidate species. Maintenance of some mollusc stocks through seeding has long been established, but systematic enhancement of crustacean fisheries particularly of lobster shows promise in North America and Europe.

Theoretically, enhancement of marine fisheries by release and recapture has a great potential to add to world fish production. Rapid though expansion is of some inland fisheries through intensification, limits will eventually be reached as the number of water bodies suitable for such development is itself limited. Furthermore, the number of species suitable for such operations is restricted. The marine environment on the other hand would seem to offer a greater area and a wider range of suitable species and thus a greater potential before natural constraints appear. However, present return rates in most cases are such that the cost–benefit ratio of stocking is insufficient to warrant it being adopted as a major approach to management. As the cost of production per stocking unit falls and as fish supplies fail to keep up with demand, causing changes in price structure, the ratio may be forced towards a more favourable balance.

1.5.2 System enhancement

Attempts to resolve limitations imposed by the natural productivity and the form of the recipient system centre around manipulation of the ecosystem, a practice which has grown not only to enhance productive capacity of selected species but also to mitigate deleterious changes to the ecosystem. Two types of action are common, interventions to increase productivity by nutrient inputs and physical modifications to the structure of the system.

Massive fertilisation programmes in natural ecosystems are not common, although fertilisation, usually from natural sources associated with polycropping systems, has been used particularly in China. It is eventually conceivable that productivity enhancement will become increasingly widespread through controlled eutrophication. Structural modifications have been practised at least since the end of the last century, when many recreational salmonid fisheries were enhanced by a wide variety of in-stream structures. Such practices have become increasingly widespread with the need to mitigate and rehabilitate heavily impacted systems (Cowx 1994). Efforts at restoration are not limited to inland systems where efforts to restore channel diversity and flood plain connectivity to rivers are increasingly common. In the marine environment, restoration of vanished *Posidonia* beds and increasing available shelter and breeding space through artificial structures such as artificial reefs are finding favour in many parts of the world.

1.6 CONCLUSIONS

A similar picture emerges in both marine and inland environments of an increasing aquaculture sector as against a generally static or declining capture fisheries sector. Declines in capture fisheries are not uniform and there are notable areas where increases are being registered. Trends in capture fisheries in both marine and inland waters point to a general failure of the policies adopted for the management of their fish stocks. They also indicate a failure to come to grips with the complex problems of containing impacts of environmentally damaging practices. The current poor state of the resource implies that substantial increases in production from traditional sources in natural waters might be hard to find, although, given the right economic and technological climate, increases in supply might be achieved through the exploitation of the marine meso-pelagics.

The rapid expansion of aquaculture in the 1970s and 1980s is probably not sustainable as an increasing number of constraints are beginning to appear to limit expansion. The hitherto fastest-growing sector economically, that of the higher value carnivorous fish and crustacean species, is likely to be limited by supplies of the fish meal upon which they depend. Aquaculture for low food chain species such as carp and tilapia presents the greatest potential. This is not being realised at present, however, because of consumer preferences and contrary socio-economic circumstances in many of the tropical areas presenting the best potential.

Enhancement of natural fisheries through culture-based systems is developing in several areas of the world and is responsible for the increases noted in inland water capture fisheries in China and parts of south east Asia. Techniques to improve productivity of inland waters are also being adopted in Africa, Latin America, Europe and North America. Extension of this approach to the ranching of marine stocks is already well advanced in Japan and, in the case of certain species, in Europe and North America. This technique presents enormous potential, but given the present economic climate and the present stage in the development of fry production technology, cost effectiveness is low.

ACKNOWLEDGEMENTS

The overview presented in this paper results from discussions within the Inland Water Resources and Aquaculture Service of the Fisheries Department of FAO. I thank my colleagues for their assistance formulating these views. In particular I express my appreciation to Jim Kapetsky, Albert Tacon, Tomi Petr and Manual Martinez for their help in reviewing the manuscript and for their constructive comments thereon.

REFERENCES

Baluyut, E.A. (1983) Stocking and introduction of fish in lakes and reservoirs of the ASEAN countries. *FAO Fisheries Technical Paper*, **236**.

Caddy, J.F. (1993) Toward a comparative evaluation of human impacts on fishery ecosystems of enclosed and semi-enclosed seas. *Reviews in Fisheries Science*, **1**(1), 57–95.

Cowx, I.G. (1994) *Rehabilitation of Freshwater Fisheries*. Fishing News Books, Blackwell Scientific Publications, Oxford.

De Silva, S.S. (1988) Reservoirs of Sri Lanka and their fisheries. *FAO Fisheries Technical Paper*, **298**, 128.

De Silva, S.S., Lin, Y., & Tang, G. (1992) Possible yield-predictive models based on morphometric characteristics and stocking rates for three groups of Chinese reservoirs. *Fisheries Research*, **13**, 369–80.

van Densen, W.L.T., Steinmetz, B., & Hughes, R.H. (1990) *Management of Freshwater Fisheries*. Proceedings of a Symposium Organized by the European Inland Fisheries Advisory Commission, Goteborg, Sweden, 31 May–3 June 1988, Pudoc, Wageningen.

Ebel, W.J., Becker, C.D., Mullan, J.W., & Raymond, H.L. (1989) The Columbia River – toward a holistic understanding. In: *Proceedings of the International Large River Symposium* (ed. D.P. Dodge). *Canadian Special Publication on Fisheries and Aquatic Science*, **106**, 205–19.

FAO (1992) *Review of the State of World Fishery Resources*, Part 1. *The Marine Resources. FAO Fisheries Circular*, **710**, Rev. 8, Part 1.

FAO (1993) *The State of Food and Agriculture 1993. FAO Agriculture Series*, **26**.

Isaksson, A. (1988) Salmon ranching: a world review. *Aquaculture*, **75**, 1–33.

Juarez-Palacios, J.R. & Olmos-Tomassini, M.E. (1992) Tilapia in capture and culture-based fisheries in Latin America. *FAO Fisheries Report*, **458**, Supplement, 244–273.

Juarez-Palacios, J.R. & Varsi, E. (1993) Avances en el manejo y aprovechamiento acuicola de embalses en América Latina y el Caribe. *FAO, Rome, GCP/RLA/102/ITA*, **8**.

Kapetsky, J.M. (1994) The potential for warm water fish-farming in Africa: a strategic assessment from a continental viewpoint. In: *Measures for Success: Meteorology and Instrumentation in Aquaculture Management* (ed. P. Keslemont), pp. 67–73. CEMAGREF Editions.

Laureti, E. (Comp) (1992) Fish and fishery products: world apparent consumption statistics based on food balance sheets (1961–1990). *FAO Fisheries Circular*, **821**, Rev. 2. Rome, FAO.

Li, S. (1988) The principles and strategies of fish culture in Chinese reservoirs. In: *Reservoir Fishery Management and Development in Asia* (ed. S.S. de Silva), pp. 214–23. IDRC, Ottawa, Canada.

Lorenzen, K. (1995) Population dynamics and management of culture-based fisheries. *Fisheries Management and Ecology*, **2**, 61–73.

Lu, X. (Comp) (1992) Fishery management approaches in small reservoirs in China. *FAO Fisheries Circular*, **854**.

Mari-Diaz, A. (1989) Pesquerías derivadas de la acuicultura en aguas interiores de Cuba. In: *Manejo y Explotación Acuicola de Embalses de Agua Dulce en América Latina*. Proyecto FAO/Italia (GCP/RLA/ 075/ITA). Brasilia D.F., Brazil, 235 p.

Moehl, J.R., Jr. & Davies, W.D. (1993) Fishery intensification in small water bodies: A review for North America. *FAO Fisheries Technical Paper* **333**.

Pavlov, D.S. & Vilenkin, B. Ya. (1989) Present state of the environment, biota and fisheries of the Volga river. In: *Proceedings of the International Large River Symposium* (ed. D.P. Dodge). *Canadian Special Publication on Fisheries and Aquatic Sciences*, **106**, 504–14.

Petr, T. (1994) Intensification of reservoir fisheries in tropical and subtropical countries. *Internationale Revue der Gesampten Hydrobiologie*, **79**, 129–36.

Petr, T. (in press) The future of fisheries in natural lakes under environmental stress in tropical Asia. Paper presented to *International Conference on Fisheries and the Environment: beyond 2000*. Kuala Lumpur, 6–9 December 1993.

UN (1991) World Population Prospects 1990. United Nations, New York.

Watanabe, T., Davy, F.B. & Nose, T. (1989) Aquaculture in Japan. In: *The Current Status of Fish Nutrition in Aquaculture* (ed. M. Takeda & T. Watanabe). *Proceedings of the Third Symposium on Feeding and Nutrition in Fish* pp. 115–129. Tilsa, Japan.

World Resources Institute (1992) *World Resources 1992–93*. Oxford University Press, Oxford.

Chapter 2
A Systems Approach to Aquaculture and Environmental Management

James F. Muir   *Institute of Aquaculture, University of Stirling, Stirling, UK*

2.1   INTRODUCTION

The world is facing increasing demands for food supply, and has a need for higher quality food resources, in which aquatic products form an important component (UNDP 1994). However, after significant growth in the last two decades, the world's capture fisheries are becoming increasingly limited by physical and biological capacity, by deteriorating environments, and by resource costs of excessive levels of exploitation. In contrast, in aquaculture, the productive inputs, processes and quality of output can be controlled to varying extents, and ownership, care and environmental responsibility may be more easily established and effectively maintained. By removing the limits to survival and productivity common to capture fishery stocks, and by effective husbandry and management, the potential for aquatic production need be limited only by the availability of simple and definable inputs such as land, water, seed, fertiliser and feeds.

According to FAO statistics (FAO 1995), aquaculture has grown steadily in recent decades (Table 2.1) and by the year 2000 farmed aquatic production might account for some 20 million tonnes per annum, some 18–25% by weight, and perhaps as much as 45% by value of the total production of fishery/aquatic products. By region, more than 80% of aquaculture production comes from Asia, less than 5% from Latin America, whilst in 1993, Africa produced only some 0.2% of global production. Although the rate of growth has been higher in developing countries (Table 2.2), much of the current investment and expansion involves more intensive forms of production, usually of high unit value species, whose returns stimulate the capital development and support the input costs. Species groups are developing at different rates, with an increasing role for fish and crustaceans, whilst seaweed and mollusc production has been fairly static. Though the production of carp species in fresh waters in fertilised or partially fed systems still dominates global output, species such as tilapia and the salmonids have shown the greatest production increases. For crustaceans, freshwater production has been relatively static, but in coastal waters the culture of penaeid shrimps has expanded significantly.

Table 2.1   Fisheries and aquaculture production, 1984–93 (10^6 t).

	1984	1986	1988	1990	1993	% annual change
Fisheries	77.3 (92.1%)	84.2 (90.7%)	87.9 (89.0%)	85.1 (87.6%)	84.7 (83.9%)	+ 1.02
Aquaculture	6.6 (7.9%)	8.6 (9.3%)	10.9 (11.0%)	12.0 (12.4%)	16.3 (16.1%)	+ 10.56
TOTAL	83.9	92.8	98.8	97.1	101.0	+ 2.08

Source: FAO (1995). Note: figures exclude mammals and seaweeds.

Table 2.2   Aquaculture production by country group (10^3 t).

Country group[1]	1984	1986	1988	1990	1992[2]	% annual growth '84–90
Developed	2761 (27.3%)	3069 (25.1%)	3366 (23.1%)	3408 (22.2%)	2600 (18.7%)	3.6
Developing:						
all species	7357 (72.7%)	9141 (74.9%)	11190 (76.9%)	11915 (77.8%)	11300 (81.3%)	8.4
fish only	3386 (33.5%)	4811 (39.4%)	6170 (42.4%)	6824 (44.5%)		12.4
TOTAL	10118	12210	14556	15323	13900	7.2

Source: FAO (1992, 1995); including aquatic plants, excluding mammals.
[1] according to standard UN categories of GDP
[2] excluding aquatic plants

As well as changes in size and output, aquaculture production methods have changed significantly, moving from traditional semi-natural approaches with low input and relatively small modifications to habitats and species distribution, towards modern intensive tank- and cage-based techniques with high and concentrated levels of input, significant targeting of species and stocks, and potentially high levels of external impact. The pace of this change has accelerated, increasing the potential for environmental consequences (GESAMP 1991; Barg 1992), though social benefit may be more difficult to discern (Bailey & Skladany 1990). Traditional methods are still used, but these too are evolving and usually intensifying in response to changes in resource availability, economic demand and technology.

2.2   AQUACULTURE, ENVIRONMENT AND RESOURCES

Aquaculture is capable of significant development, and may in at least some circumstances help to address increasing problems of food supply and food security. However, its dependence on natural resources, and its potential for placing greater demands on these, may place it in direct competition and possible conflict with other demands, whether utilitarian or orientated around conservation or ethical issues. Internationally, rising demands for natural resources for economic growth and productive development have been accompanied by increasing public sensitivities to environmental impact. The 1992 United Nations Commission for Environment and Development (UNCED) Conference brought these issues to wider public attention and political focus, resulting in a series of commitments for environmental protection and sustainable development. These have highlighted the need for better management of natural resource-sensitive processes such as aquaculture, and are encouraging the process of bringing the externalities of actual or potential impacts into direct environmental costs (Folke & Kautsky 1992) and into environmental and resource management policy.

Together with increasing commercial pressures, these trends are bringing about a growing need for efficiency within the aquaculture sector as a whole, in which inputs need to be more carefully applied and managed. The main issues and key environmental problems (Håkanson *et al.* 1988; GESAMP 1991) associated with these factors are shown in Table 2.3. After a considerable 'honeymoon period' in which the precarious yet promising technical and financial aspects of the developing aquaculture sector encouraged widespread support in terms of policy and investment, the industry started to expand, moving into previously undeveloped areas, and competing for resources. As a result, an increasing range of actual or potential impacts was observed. Much concern has involved the nutrient outputs from aquaculture, as

Table 2.3 Environmental problems associated with aquaculture.

Problem area	Nature of problem
Waste and nutrient loadings	Outputs of solids, N, P, vitamins, minerals, husbandry/disease chemicals, antibiotics; impacts of waste materials on the adjacent benthos and the water column; on species/community diversity, quality indices, stimulation of blooms.
Water exchange	In intensive land-based systems, or flushing through freshwater or marine cage or enclosures; quantities required, effects of abstraction, dilution with 'low grade' wastes, at concentrations sufficient to diminish measured quality, but too low for simple treatment.
Degradation of terrestrial environments	In coastal areas, caused by salinisation of soils, affecting adjacent agricultural practices, coastal fringes; excessive clearance of mangroves and protective cover.
Escaped stocks	From damaged systems, or through flooding, damaged or ineffective discharge screens; risks of competition with/genetic contamination of local stocks, disease transmission, directly or indirectly reduced biodiversity.
Predation by conservation sensitive species	Causing damage, loss, stress-related disease to farmed stocks, requiring controls without compromising conservation interests.
Social/amenity disturbance	Visual, noise, activity disruptions.

these have been the most obvious and usually most directly measurable. Requirements have usually been met by supplying copious quantities of water, discharging to receiving systems having a high absorptive capacity. A potentially more serious concern is that of biodiversity (Beveridge *et al.* 1994).

At the same time the increasing intensiveness of many forms of modern aquaculture, using species high in the natural food chain, and the resource demands implicit in this, led to analogies with the most intensive agricultural production practices (Leach 1976). The realisation grew that the most actively growing sectors were highly inefficient in ecological energetic terms, and exhibited many of the undesirable characteristics of resource concentration, both globally, from poorer to richer countries, and locally, from surrounding areas towards the sites of production. Moreover, far from being a widespread benefactor in terms of local jobs and economic multipliers, many of the more commercially efficient production units became increasingly mechanised, with reduced numbers of jobs, of low skill level, and offered little if any linkage with local communities. Allied to the significant dependence of many of the carnivorous species on high quality feed ingredients, particularly marine fish meals and oils (Tacon 1994), the future acceptability of many forms of aquaculture and its potential for development could be cast in serious doubt.

Limitations in resource quantity or environmental capacity may be tolerable in the short term, but may quickly lead to unacceptable effects (Beveridge 1984). A wide range of problems have arisen when loadings have exceeded environmental capacity (Pullin *et al.* 1993). Chua (1993) and Briggs (1993) noted the problems of the Taiwanese shrimp industry in 1988, 1992 and 1993, and the collapse of production in central Thailand in 1989–90, with an estimated loss of some $US30 million and the dereliction of some 45 000 ha of land. Problems have also been noted in the Philippines, Indonesia, Sri Lanka, and in China (Chua 1993), where production has also been heavily affected. The need to supply such demands, the increasing need to 'fallow' sites, i.e. move production to allow sites to recover from loadings, and increasing management restrictions, provide serious constraint for continuation or development (Jensen 1993).

As a consequence of the rising concern for the specific impacts of aquaculture, and the broader realisation of the potential range and possible gravity of effect, a steady ground shift of public opinion and political pressure has in many areas of the world resulted in a more negative and regulatory approach to aquaculture, an increasing suspicion of development, and an increasing demand for pre-emptive control (Muir & Baird 1991). This might typically require a precautionary or even 'zero-tolerance' approach to existing and intended development, implying anticipatory understanding of processes and risks, and provision to manage or compensate for even very low levels of environmental effect. While such a response can be understandable, and may be seen as a short-term corrective, there is a risk of exaggerating the severity of aquaculture's impacts, or focusing on the wrong issues. Instead of encouraging good practice, sound husbandry and rational development, reactive political and regulatory decisions could unnecessarily restrict the potential of aquaculture or produce the wrong development choices, and may allow other more environmentally-damaging activities to remain unchecked. This can be a particular consequence of 'single-issue' approaches to environmental matters, and is a common result of the equivalent campaigning, which characterises a certain part of environmental debate.

It is therefore important that the correct choices are made concerning environmental quality, resource use and economic output (Pimentel & Pimentel 1979; World Bank 1992), that aquaculture is allowed to take its part effectively in meeting future needs, and that its promoters be given the appropriate market and policy signals to develop more effective and acceptable means of production. In broad terms, we might aim for sustainability, and seek to define sustainable forms of aquaculture. However, this cannot be done by considering the aquaculture sector and its activities in isolation; rather it is important to attempt to understand activities and processes within a wider systems context, where aquaculture may occupy at most only a partial role in locational demands, resource use and socio-economic significance.

By considering the nature, presence and inter-relationships of aquaculture within a systems context, involving a far broader range of issues such as environmental capacity, land-use processes, social preference and economic choice, it may be possible to place it more rationally in the practical policy sphere, and deliver the appropriate longer-term benefits. The following sections attempt to review the concepts of systems perspectives concerning aquaculture, to consider the implications for resources and development, and to offer outlines for ways in which these areas of understanding might be applied to support efficient and environmentally-benign development. An initial step to doing so would be to review the concepts and constructs of sustainability, as this might provide the framework and the specific objectives for which a systems perspective might be developed.

2.3   THE CHALLENGES OF SUSTAINABILITY

If a workable approach is to be proposed for aquaculture and its development, in which the interplays between resource use, economic and other outputs are to be expressed, the concept of sustainability must be the first point of consideration. The notion of sustainability, and associated terms such as 'sustainable development', are now widely used in discussion on economic development, in issues of conservation and in the social and economic management of production systems, especially those closely tied to natural resources, and are increasingly applied to aquaculture. In particular there is a wider public awareness of conflicts between the perceived needs for conservation and environmental protection, and the needs of agricultural or industrial development and economic growth, where capital wealth could be seen to be created at the expense of 'natural wealth'. Though the initial concepts have a longer history, the first major recognition of the broader goal of sustainable development was presented by the Brundtland Commission (WCED 1987):

'Sustainable development is development that meets the needs of the present without compromising the ability of future generations to meet their own needs'

2.3.1 Definitions of sustainability

By the time of the UNCED 'Earth Summit' in Rio de Janeiro in 1992, a major political recognition was taking place concerning changes in the perceived relationships between the human economy and the environment. Sustainability or sustainable development is now commonly included within the mission statements of conservation groups, governments, national and international agencies, and commercial enterprises. Jacobs *et al.* (1987) noted sustainable development as 'a reaction against the *laissez faire* economic theory that considered living resources as free goods, external to the development process, essentially infinite and inexhaustible'. Five elements were noted:

- integration of conservation and development,
- satisfaction of basic human needs,
- achievement of equity and social justice,
- provision for social self-determination and cultural diversity, and
- maintenance of ecological integrity.

According to Pearce (1991) most definitions suggest as core issues:

- Futurity: a concern for the well-being of future generations (inter-generational welfare),
- Equity: concern for future generations should extend to equity for the current generation (intra-generational welfare),
- Environment: greater emphasis on the environment.

The definition of sustainable development given in the Brundtland Report is generally the most commonly quoted. However, though it is increasingly being urged on every productive sector, including aquaculture, its implications and consequences are very rarely spelled out beyond this thematic level. As Gow (1992) noted 'sustainability is like happiness – everyone believes in it and everyone has a different definition'. This imprecision may indeed strengthen its role as a catalyst for change. Reid (1993) described comments by Jacobs *et al.* (1987), noting that by fudging terminology it may have been more supportable politically, the starting point for the necessary process of change. Peet (1992) quoted de Vries as saying that 'sustainability is not something to be defined, but to be declared. It is an ethical guiding principle'.

2.3.2 Levels of sustainability

Clearly, sustainable development can mean different things in different contexts, and though most involve the three elements above, the priorities may vary. As Table 2.4 indicates, there may be different levels of rigour, depending on the degree to which the future may be expected to look after itself. The range extends from the *laissez-faire* extreme of technocentric faith, in which scientific and technological progress can overcome any environmental cost associated with development, and provide a range of substitute goods for those that are depleted or can no longer be provided (the 'technical fix' scenario), to the other extreme 'environmental doom' scenario, which assumes any human intervention or modification to be potentially disastrous unless it can be rigorously proved otherwise. The implications for aquaculture, water resources, their ecosystems and their management are also drawn. It is important to note that these descriptions differ not simply in the technical choices involved, but in the moral and ethical underpinning and the implications for societal response. One can also note, as in other areas of ethical/socio-political choice, that extreme libertarian approaches can be identified at each end of the spectrum. The logical extension of complete ecocentric rights is *laissez-faire*, unmanaged wilderness.

To approach policy development for aquaculture at a simple level, sustainability may merely mean that an activity can continue or a resource be available for a definable term, can be self-sustaining (e.g. an externally-funded development project) and is perhaps not associated with or dependent upon rapidly depleting inputs, particularly if non-renewable. Thus, at a practical level (SNH 1993) 'levels of renewable resources will be maintained in the course of exploitation, with resulting benefit for the social and economic framework, while enhancing, or at least avoiding damage to the environment' and 'profitable production, sustaining rural communities and structures, and protecting soil, water and air quality, and maintaining or enhancing the landscape, biodiversity and wildlife conservation'.

In a more fundamental and wide-ranging application, however, sustainability would require no diminution of future potential, and provision of intergenerational equity in any or all elements of the geosphere. It may then focus quantitatively on the issue of maintaining and possibly increasing the world's 'capital stock' – defined basically as the quantity of physical and biological resources, but also extendable qualitatively, and beyond immediately utilitarian matters, to include e.g. human skills, technological potential, sources of satisfaction, etc. As Fig. 2.1 indicates, three areas of stock can be defined, all of which may contribute to present or future well-being. If the societal well-being can be defined as the total of these capitals, such a structure might then allow for depletion of one or more elements if matched by gain in other areas. In other words, there may be tradeability between environmental goods and economic output, possibly even if these are damaging, yet still allowing the overall criterion for sustainability to be met. This is the overall theme on which a

Table 2.4 Sustainability levels, water resources and aquaculture.

Sustainability Level	Very weak	Weak	Strong	Very strong
Conceptual description	Technocentric/ cornucopian	Technocentric/ accommodating	Ecocentric/ communalist	Ecocentric/deep ecology
Resource approach	Exploitative growth-orientated	Conservationist/ managerial	Preservationist	Extreme preservationist
Economy approach	Free market	Guided free 'green' markets	Regulated steady state	Heavily regulated
Management	GNP goals, free markets, technical progress provide infinite substitution possibilities	Green accounting modified GNP. Limited substitution. Constant capital	Zero economic and population growth. No scale increase. Systems perspective	Reduced scale of economy and population
Ethics	Rights of contemporary individual humans. Instrumental values	Caring for others temporally and spatially. Instrumental values	Collective interests, ecosystem value, functions and services secondary	Bioethics – other species' rights. Intrinsic values
Water resource implications	Market driven allocation and development decisions. Free pricing and price/ quality choice	Value/quality/price decisions within managed framework. Preservation of basic access rights	Control on use/ suitability, pressure to upgrade quality throughout, limit deleterious use	Strong control on use/suitability quality paramount, irreversible high entropy change controlled/ prohibited
Aquaculture development implications	Market-driven decisions, purchase of resources according to means, environmental control developed/ applied only for management gain	Lightly regulated according to basic criteria, but generally open access and markets, environmental control may be required	Broadly regulated across range of criteria, with EIA requirements, proof of no damage, rehabilitation clauses, restrictions	Presumption against most forms of aquaculture, except low input/ impact, with strong precautionary emphasis
Ecosystem implications	Valued/protected if identified as contributing to water value and/or definable recreational value	Managed to satisfy wide range of costed/uncosted factors. Responsive to social demands. Moderately precautionary	Precautionary approach, presumption against change, limit effects of use. Rehabilitation Ecosystem quality objectives	Absolute precautionary; presumption against existing changes, ecosystem quality objectives and rehabilitation paramount

Developed after Pearce (1991)

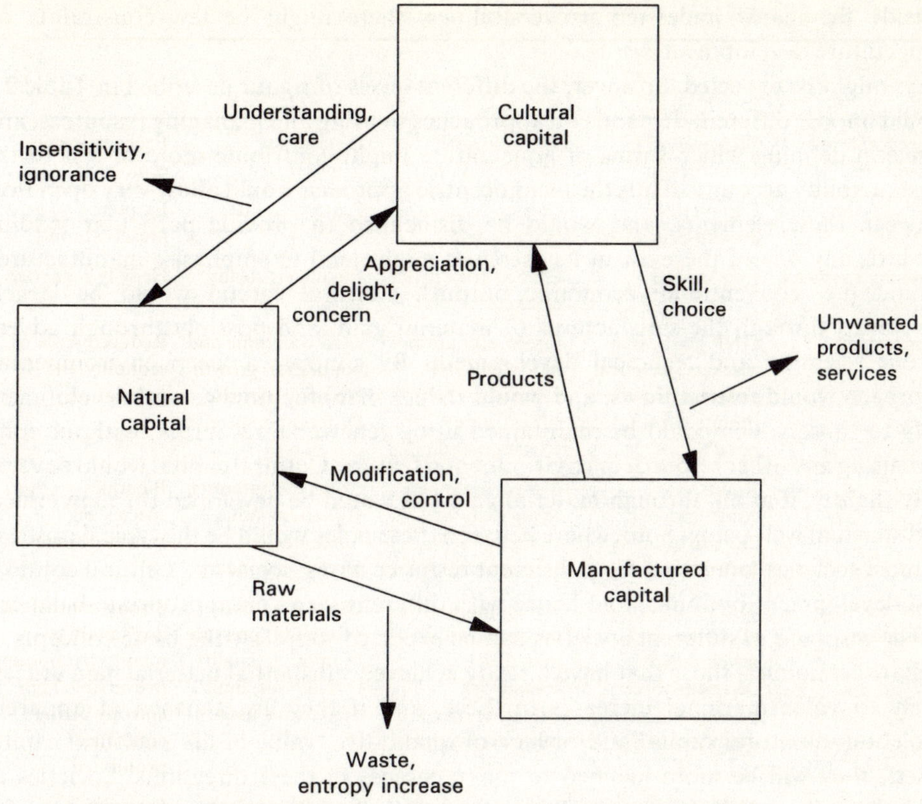

Fig. 2.1   Sustainability relationships.

systems approach can be developed, and in which the role and potential of aquaculture might be considered.

As will be discussed later, this is a system view in its own right, and hence it can be viewed as the overall guiding framework for other systems approaches. Before reviewing the nature and consequences of systems and their use in development, it is useful to consider some of the questions and implications arising from this sustainability structure. While the system described in Fig. 2.1 offers a useful general perspective, there can be substantial methodological problems in defining common 'units of currency' between and within each area of capital. In the case of natural capital in particular, various constituents might be identified as 'non-tradeable', i.e. their consumption or transfer would be environmentally and/or socially unacceptable. The point is, however, that the overall approach allows non-renewable resources to be used in development, e.g. if by doing so it increases the net benefit stream or improves societal ability to generate benefits from other forms of capital through mechanisms such as technological development. This would imply that, in general terms at least, outside the major undesired irreversibilities, there might be few constraints on aquaculture development *per se*.

As might be expected, however, the different levels of rigour described in Table 2.4 would impose different demands on approaches to using and managing resources, and hence in defining which forms of aquaculture might contribute more or less to the 'sustainability account'. Thus the technocentric approach would allow very open flow between these elements, and would be disinclined to provide particular loading towards any one of these, or in its extremes might tend to emphasise manufactured capital (i.e. conventional economic output). Cultural capital would be largely developed through the satisfactions of material gain, and possibly through adventurous scientific and technical development. By contrast a 'deep environmental' approach would restrict flows, and would reduce manufactured capital development only to those which could be maintained using renewable resources, without compromising any other resource or environmental feature. Cultural capital would develop only slightly, if at all, through material gain but would be developed through ethical and spiritual well-being. Somewhere between these poles would be the typical position of most societies, operating from different resource bases, social and cultural contexts and development outlooks, and hence with different views on appropriate balances.

The response of different societies can be expected to follow the basic concepts of welfare economics; those that have already achieved substantial material gain are less likely to value marginal increases in these, and if they are alarmed at apparent depletion in natural capital and/or lack of qualitative 'value of life', cultural capital assets, they will be more inclined to make choices in these directions. Societies or decision-making individuals who have achieved little material gain are more likely to purchase these at the expense of natural capital. This pattern can only be observed in general terms for aquaculture development, in which the most prosperous nations are less inclined to trade their own natural capital for further output (but are normally quite well prepared to purchase products from elsewhere), while poorer countries are all too willing to do so, and will trade their natural resources for domestic production and export earning. However, the argument may not be so simple as this; societies with poorly developed cultural and ethical capital, i.e. dissociated and alienated societies, are more likely to exploit their own and other resources, whereas those with a strong and cohesive ethic may seek to retain conservationist approaches in spite of low levels of material wealth.

The use of this structure, and the associated arguments, can be developed further in terms of the relationships between macro-level (e.g. global, national, or ecosystem) sustainability and that at the specific level (e.g. locality, community, or project-based). For example, the scheme above suggests that larger-scale sustainability does not require smaller-scale sustainability to be universally attained. That is, it could be feasible to accept a series of definably non-sustainable activities, such as highly intensive aquaculture, as long as their net deficit, e.g. depletion of non-renewable energy, longer-term qualitative and quantitative diminution of fish stocks, could be compensated by other activities, e.g. use of aquaculture for rehabilitating damaged environments, or indeed any other activity which had a suitably positive effect.

The trade could be done at a range of levels, for example at the enterprise or at the level of the national economy. This is the basis for various 'environmental compensation' approaches. Clearly, this holds only if the non-sustainable activities do not endanger those critical non-tradeable natural resource areas which might compromise the sustainability of the rest of the system. Very localised and subject-specific interpretations of sustainability might also be noted, i.e. the continuation of a particular set of benefits, activities or characteristics deemed to be valuable by specific sectors of the society. While in some cases these may be incidentally related to general sustainability, there is not necessarily any specific functional linkage, and indeed the demands to maintain such a set of benefits may counteract the general good. It is clearly important to recognise the different usages of the term, to avoid confusion and misplaced objectives.

2.3.3   Operating a sustainability system

In aquaculture, as in most areas of human activity, there are quite direct issues concerning the efficiency of resource use and the benefits derived. However, there are also more indirect, possibly more subtle, issues, such as e.g. potential non-reversible effects of present-day actions, not all of which may be apparent at the time of their commission. The precautionary principle as described in Table 2.4 is often invoked as a means of avoiding such risks. In outline, this would require that if any area of doubt concerned the consequences (usually environmental, but also social or economic) of a particular development, it should not proceed, or at least should contain compensatory mechanisms for the worst-case outcome. This is by no means widely accepted (Pearce 1991), even if recognised, as economic imperatives of present-day demands may not allow deferral of key decisions to provide food, income, and other benefits. In fact one of the little-observed implications of the sustainability argument is that, unless there are obvious future consequences, natural resources should be employed for material development in the present generation, to meet intra-generational, social equity needs. That is, development in areas such as aquaculture could proceed, if there is a material need for them, i.e. supplying food and improving food security.

There are many methodological issues involved in defining and prescribing sustainability and its various elements, and a rapidly growing literature (e.g. Pearce 1991; Lele 1991; Chambers & Conway 1992). There are difficult questions concerning the comparability of environment functions and values, definitions of development with respect to material, social and cultural well-being, and balances between intra- and inter-generational equity. There is also the issue of 'interspecies equity' as demanded by strong sustainability, and a range of other issues of conceptual and practical import. For natural capital in particular, various constituents might be identified as non-tradeable, i.e. their consumption or transfer would be environmentally and/or socially unacceptable. However, if this can be effectively defined, it does allow non-renewable resources to be used in development, e.g. if it increases the net benefit stream, or improves societal ability to generate benefits. In practice, a range of mechanisms has developed, and is evolving further, including various planning and regulatory controls, *ad hoc* or policy-based, and methodologies such as EIA (environmental impact assessment – Shopley & Fuggle 1984).

Given the importance of social equity and the concepts of social involvement, or stakeholding, with an informed polity, as being central to deciding the balances between natural, material and cultural capital, there has also been increasing emphasis on participatory approaches, particularly in developing countries (Chambers & Conway 1992). Many of these methods and approaches have been applied to aquaculture development. However, to date, few of these procedures have been specifically organised to deal with sustainability *per se*; rather they are concerned with resolving present-day conflicts of interest. In consequence, though some of the system relationships described in this basic sustainability perspective may be explored in passing, they are not done as a matter of deliberate intent, with more comprehensive and potentially long-lasting solutions in mind.

This section of the review has explained the basic format for sustainability and the choices therein. It is described loosely as a system, in that components and relationships can be defined, and it can be proposed that such a system could define the overall objectives in which aquaculture could be developed. It can also be considered that, if effectively applied, sustainability approaches would incorporate environmental concern to an appropriate level, and that this could provide the framework for development and environmental choice. The next section examines in more detail the nature and characteristics of systems, and considers how they may be applied to aquaculture and its development.

2.4   A SYSTEMS PERSPECTIVE

Before attempting to describe aquaculture from a systems perspective, it is useful to offer some primary definitions. Firstly, FAO (1995) defines aquaculture as:

'the farming of aquatic organisms, including fish, molluscs, crustaceans and aquatic plants. Farming implies some form of intervention in the rearing process to enhance production, such as regular stocking, feeding, protection from predators, etc. Farming also implies individual or corporate ownership of the stock being cultivated. For statistical purposes, aquatic organisms that are harvested by an individual or corporate body that has owned them throughout their rearing period contribute to aquaculture, whilst aquatic organisms that are exploitable by the public as common property resources, with or without appropriate licences, are the harvest of fisheries'.

This basic definition, though as stated, primarily developed for statistical purposes, essential to distinguish aquaculture from capture fisheries, already holds within it various elements that can be recognised as necessary contributions to the activity and its characteristics, i.e. that contribute to the system that is aquaculture production. It involves social and economic organisation, biological (and hence ecological/environmental) management, the application of external resources, and by implication, the use of physical arrangement, e.g. for containment and/or protection. It is in effect a system of production, but what are the implications of using the word system to describe this, and can it confer other levels of understanding and use?

Amongst a range of broadly similar descriptions, *Webster's Dictionary* (9th edn) defines 'system' as a 'regularly interacting or interdependent group of items forming a unified whole'. In this way an aquaculture enterprise can be said to operate as a system, composed for example of management framework, physical facilities, stock, water, pathogens, predators, feeds, labour, market, these units forming part of its structure. The idea of interaction and interconnection implies that components within the system can influence one another, through a series of inputs and outputs. In this basic example decreased water flow will influence stock, resulting in changes in their resistance to pathogens, reduced feed intake, increased wastes in water flow, and reducing the quantity and quality of output to the market. Similar series of influence can be described in the other systems involved in aquaculture production. Classically, these can proceed in a linear direction, or there may be one or more points of feedback, i.e. where influences return to affect the initial impulse or disturbance.

The behaviour of a system will therefore be determined by the nature of its structure, and by the nature and size of the disturbances or perturbations it experiences. The interacting and interdependent parts of a system are linked by exchanges of energy, matter and/or information. Tentatively, aquaculture systems might then be described as 'the interdependent elements of environment, physical structure, rearing process, resource use and socio-economic context which form the entity from which farmed aquatic organisms are produced'.

2.4.1 System characteristics

The boundaries chosen to define the system are critical in providing explanatory sense for that system, and describe its context and its nature. The primary concept is that the elements inside the boundary form the system structure, and interact to provide the system characteristics. This assemblage is then distinct from external elements, which act on the system, but do so primarily in a simple linear manner and are not involved in the process of the system itself. For a given context, the boundaries can be placed at a range of levels. For aquaculture these may, for example, be drawn at the level of the organism, the production unit, the economic enterprise, or at the national sectoral level. This indicates the potential for systems to be linked together, and indeed for hierarchies of systems to be described, with one system overlapping another, or 'nesting' inside a larger system.

In classical systems analysis, boundaries are chosen to minimise interactions with other systems and levels, providing the means to break down complex systems into components, subsystems, and hence to understand better the structural and functional relationships within defined limits and between systems. In practical terms, without too complex analysis, systems boundaries can often be established in association with physical entities, geographical borders or organisational structures. They may be adjusted outwards to bring in external elements that affect the system in more fundamental and complex ways, or inwards to exclude elements that are not intimately involved in the behaviour of the system.

A fundamental feature of any systems view is then the understanding of interconnections and hierarchies. These have been alluded to in the previous discussion on sustainability, and can be more explicitly connected together using a systems structure. Table 2.5 provides a summary of typical scale hierarchies as they may apply in the aquaculture/water resource context, with basic factors classified in broad terms as physio-chemical, bio-ecological, socio-economic and production-related. Though scale dimensions are not closely specified, the classifications used give some indication. Thus 'geo'-level systems are essentially defined by the scale of the earth and its major features; macro-level systems correspond to major sub-elements of these (typically related to areas of 10^3–10^6 km^2 and populations of 10^6–10^9 individuals at the level of humans or mammals or many scales of magnitude greater for complete species inventories); median-scale systems correspond to areas of 10^{-1}–10^4 km^2 and 10^3–10^6 individuals; small-scale systems correspond to areas of 10^{-3}–1 km^2 and 10–10^5 individuals; and smaller-scale systems involve progressively smaller areas and populations (or significantly larger numbers of far smaller-scale organisms).

Clearly, these scales link together, both vertically in the sense that larger scale units contain smaller scale units and are ultimately defined and characterised by them, and horizontally in that the system classifications link together (though there is some range in scale in the different areas). It is important, however, to understand the capability of each system, at whatever scale, to be seen as a single functional unit, and thereby to be able to accept inputs from other systems, process them, and provide outputs in turn.

The processes within a defined system are the features which determine its behaviour and response to internal and external influence and perturbation. Some of these are explicit and can be defined by causal links, e.g. physical, thermodynamic, chemical, stoichiometric, biochemical, which may be linear, function-law-related or at least relatively simple and deterministic. Others can be broadly described, e.g. transport, metabolic, behavioural, as a combination of various process subcomponents, and are to varying degrees explicit and deterministic. Others, such as ecological or social processes, though recognisable as features of a system, are less easily defined and characterised, and may be described only at the levels of direction

Table 2.5   Scale hierarchies for systems related to aquaculture and water resources.

Scale level	Physico-chemical	Bioecological	Socioeconomic	Production
Geo	Oceanic/climate system. Hydrological cycle. Ocean biogeochemical cycle	Biosphere. Oceanic ecosystem. Global species population.	Global population. Global economy. International agency.	Global food production. Global fishery sector. Global aquaculture.
Macro	Catchment/watershed. Coastal zone. Large aquifer. Large island or archipelago.	Major biomass. Large scale species assemblage.	Region/large nation. National political structure. Major religion/ethic. Major communications.	Regional/national food, fishery, aquaculture sectors. Species/system sub-sector. Major producer group.
Median	Main river section. Lake, reservoir. Smaller aquifer. Coastal segment. Small island.	Community groups. Populations. General environmental change/impact.	Small nation. Governmental institution. Large corporation. Generic market.	Producer organisation. Production entity. Production site. Integrated equipment/management assembly.
Small	Stream, bank, bed unit. Lake compartment/mixing cell. Sediment area.	Family. Inter-species. Specific environmental impact.	Small business. Specific market. Club, association.	Holding unit/small site. Equipment item. Stock group.
Micro	Local mixing cell.	Individual.	Family.	Individual stock, feed. Equipment sub-assembly.
Nano	Sediment particle. Molecular process.	Cell.	Individual.	Feed particles. Equipment parts.

of response inside the overall 'envelope' within which the response is defined. Table 2.6 shows the typical range of processes that may be relevant to aquaculture, water resources and their sustainability. As with the previous table, these are broadly grouped into physicochemical, bio-ecological, socio-economic and production-related.

2.4.2   Closed and open systems

Distinctions can also be drawn between closed and open systems. In the first of these, matter, energy or information neither enters nor leaves the system, but is conserved within its boundaries. Such closed systems are typically linear, ordered and predictable, and tend to be characterised by small inputs yielding small results. These

Table 2.6 Examples of system processes.

System process/type	Processes	Characteristics
Physio/chemical		
Fluvial.	Hydraulic, hydrodynamic, particle transport, solution, interception.	Many definable statistically, with empirically derived relationships, some related to engineering hydraulics.
Lake and wetlands.	Hydraulic, thermal, sedimentation, solution, interface, adsorption.	Less clearly defined as total systems, though individual components can be described; stochastic approaches can be used.
Ionic transport.	Solution, diffusion, mass transfer, adsorption.	Most definable according to basic relationships, but more complex interactions difficult to describe.
Bioecological		
Lake ecology.	Physical/chemical as above, nutrient transfer, growth, competition, predation, community dynamics.	Some components definable, but most in only generalised terms, with limited management predictability.
Fishery production.	As above, growth, metabolism, behavioural, disease, genetic selection, habitat interaction.	Some definable in general terms, but difficult to specify at detailed level.
Aquaculture.	As above, waste output, environmental control.	Better controlled conditions improve level of definition.
Socio-economic		
Political.	Voting, opinion formation, media response, personal and collective competition.	Can be described and characterised, few explicit mechanisms apart from basic procedures.
Organisational.	Objective targeting, business management, market and economic response, competition, regulation.	Can be described, characterised, apart from financial and economic structures, few explicit mechanisms.
Ownership.	Wealth creation, inheritance, common aims, competition, regulation.	Few explicit mechanisms apart from financial and economic.
Information.	Acquisition, interpretation, dissemination, dialectic, competition, regulation.	Few explicit mechanisms, though, e.g. market response and opinion formation can be analysed.
Production-related		
Water distribution and supply.	Hydraulic, physical, chemical, microbiological control system.	Most aspects definable, at least empirically, based on engineering design approaches.
Pond production.	Hydraulic, thermal, sedimentation, mixing, nutrient transfer, growth, competition, community dynamics.	Some components definable, but most in only generalised terms; limited management predictability.
Cage production.	Hydraulic/hydrodynamic, structural, behavioural, diffusion, sedimentation, particle transport.	Some definable empirically, moderate management and design potential.
Waste treatment.	Hydraulic, physical, chemical, microbiological, control system.	Most definable at least empirically, though biological treatment systems much less so.

Developed from Muir (1994)

systems are stable, exhibit equilibrium behaviour, and act according to Newtonian mechanics, wherein the behaviour of objects is predictable from a set of laws from known initial conditions. Simple examples would include electro-mechanical devices and engineered structures, and may be included in the range of operational systems within an aquaculture unit. However, their very simplicity and predictability renders them less critical (within their defined operating limits) to the effectiveness of the systems in which they are an element. In contrast, in open systems, matter, energy and information may both enter and leave the system at various points in time and space.

Most real-world systems, including those involved in social and economic mechanisms, in ecological processes and in complex activities such as aquaculture, are open, exchanging energy, matter and information with the environment. These open systems are characterised by non-linearity, disorder and unpredictability, where small inputs may yield large results or they may not. They may exhibit a range of behaviours, ranging from stability to oscillation and chaos, and this may be difficult or impossible to predict, even with closely specified initial conditions. As open systems become more complex in terms of structure, their behaviour becomes more difficult to predict, even in general terms.

Though the condition of equilibrium is possible in open systems, it is only one of a number of potential behaviours. In these cases, equilibrium describes a dynamic, meta-stable state, where interacting components maintain a given structure for a given time period. For such systems, stability (the ability to conserve essential properties of structure and function) can be described as local or global. Locally-stable systems exhibit stability within normal limits of environmental or other variability. However, they may not exhibit global stability, the ability to return to a normal state and following severe perturbation. The use of terms such as normal and severe indicates the arbitrariness involved, and hence the need to define these for particular types of system.

In considering the wider issues of sustainability, the role and effects of natural resource-based activities such as aquaculture, the conservation of ecological systems and aims to preserve biodiversity, more complex systems are often argued to promote stability. However, these arguments are by no means clear. Nonetheless, subject to such caveats, Holling *et al.* (1994) noted that ecosystems with high biodiversity appear to show enhanced global stability following environmental perturbations, evidence in favour of the 'complexity/biodiversity enhances stability' argument. To extend these perspectives further, Table 2.7 illustrates a range of systems related to the 'sustainability in aquaculture and water resources' issue, and shows their typical or expected behaviour under different conditions.

In its overall approach and application, systems science focuses on function and links, rather than individual organisms and processes. The sustainability of human managed production process, or systems, therefore relates to the requirements for external inputs and the productions of outputs, and how these interact with associated and enveloping systems. The widest system boundaries for practical purposes

Table 2.7   Systems: responses to perturbation.

System	(1) Regulated (local) response	(2) Stressed (global) response	(3) Unregulated response
Physico-chemical.			
Ionic.	Equilibrium processes.	Accumulation, equilibrium shift.	Release.
Particulate.	Stable/balanced movement/transport.	Accumulation/erosion amplitude change.	'Cascading', consequential change.
Structural.	Resistance/control, component balance.	Tolerance limits reached, damage, impairment of non-critical elements.	Failure, disintegration.
Multicomponent.	Stable response, processes.	Stable, higher magnitude responses.	Disintegration.
Bioecological			
Individual.	Homeostasis, behaviour response.	Susceptibility, limited scope, competitiveness, risk to reproductive success.	Mortality, reproductive failure
Species.	Genetic diversity, population resilience.	Reduced diversity, population limits.	Decline. Extinction.
Community.	Biodiversity, plasticity of response, resilience.	Reduced community diversity, elements at risk.	Disintegration, species dominance.
Socioeconomic			
Individual.	Equilibrium, stable relationships, health, responsibility.	Instability, fear, irrationality, health decline.	Breakdown, social dysfunction, mortality, suicide.
Community.	Population balance, social interaction, responsibility, good health, economic stability.	Imbalances, social diversion, increasing anomie, disease transfer, economic instability.	Conflict, aggression, disruption, economic breakdown, murder.
Society.	As above, political maturity, balance.	As above, political instability.	As above, political conflict, war.
Production-related			
Stock unit.	Homeostatic, behaviour response health, growth, reproduction.	Stress, increased susceptibility, growth, reproduction impaired.	Disease, reproductive failure, toxicity, stock loss.
Holding/production group.	Integrity, behavioural, structural balance, reliability.	Structural, functional stress, reduced safety margins.	Disintegration, loss of control, escapes, system failure.
Commercial entity.	Cohesion, management control, reliability, profit, development.	Commercial instability, risk, work/management overload.	Work/management failure, financial loss, bankruptcy.
Production subsector, producer group.	Identity, cohesion, strategic competence, development.	Reduced cohesion, separation development uncertainty.	Disintegration, internal division, breakdown of function, support.

Developed from Muir, 1994. The relationship between the three response states depends on the chronology and magnitude of the stressors. Reversibility between (1) and (2) is usually possible, but not between (2) and (3). Vertical linkages can also be noted 'down' the table. Thus small scale failure (3) levels can be accommodated within (1) and (2) levels at a greater scale, but may trigger (3) response at the greater scale.

are set by the globe, which is essentially a closed system, the sustainability of which relies on continued energy flux from the sun to drive the internal processes and transformations which make up life on earth. Within this system, all sub-systems are essentially open to a greater or less extent (connecting flows of energy, matter and information with other systems), though they can be acted upon by designed systems which may be more closed in nature. From this, sustainability can be defined as the ability of a chosen system to persist in a given form (in terms of structure and behaviour) for a specified time period according to defined criteria.

Thus for aquaculture, at the level of the production unit and its methods of production, the producer entity, or the production subsector, various system characteristics can be defined and required to continue, both for existing activities and for future activities of the particular type. This area of definition can be placed inside the wider system of 'capital exchange' shown in Fig. 2.1, in the sense that features of natural, economic and cultural capital might continue to be definable as they change. The precise limits within which system structure and behaviour can change, and yet still retain the essence of the system, will be defined by scientific, social and cultural contexts, which are themselves subject to processes of change. The sustainability of the system is also determined by the boundaries that have been defined, and, as discussed elsewhere, criteria at other boundary levels may take precedence, depending on the choices made and the trade-offs agreed. However, for any system, multiple observers (stakeholders) can be identified, each with their own views and objectives concerning the future development of the system (Chambers & Conway 1992). In this sense, sustainability is a 'soft' system concept, which outside the relationships and processes that are scientifically and technologically grounded, must ultimately be defined by competition or consensus (via whatever social or political process).

2.5 SYSTEMS AND AQUACULTURE

The previous sections have considered the requirements for sustainability and have described the ways in which a systems perspective might be used to describe and understand the role of a developing production sector such as aquaculture. This is now taken further, to consider the specific levels at which systems descriptions might be applied. The previous tables have shown examples of different system structures and hierarchies associated with aquaculture and water resource management. To summarise, Table 2.8 provides an overview of the key issues that can be described in such a systems context, ordered in a broadly hierarchical context.

This order of systems can be seen to offer an organisational and explanatory structure for the way in which aquaculture and its features and inter-relationships may be considered within the system for resource allocation, and within the framework of choice between environment, development and social/cultural gain. Thus from the 'top-down' strategic direction, the higher-level cultural, ethical, social and political systems, can be derived the knowledge and attitudes that allow techniques to

Table 2.8   Systems features associated with aquaculture and water resource management.

System level/context	Significance/features
Ethical, cultural.	Social value structures in which decisions may be made, concerning e.g. consumption, attitudes to sustainability, conservation, tradeoffs; also attitudes to development, equity.
Socio-political, economic.	Systems in which social wishes articulated, expressed, current and shorter-term value defined, resource allocation choices made.
Climatological, agro-ecological.	Physically defined systems in which production located, defining environmental, structural, operational parameters, context for nutrient, energy exchanges.
Production, farming (FSR, FSD).	Physically and/or socially-defined local context for production unit, local resources, family, community activities, choices, interchange of materials, energy, financial/other value.
Resource use, mass-balance.	Physically, chemically, biologically defined; context for describing nutrient, energy exchanges, quality transformations, environmental capacity.
Bio, eco.	Context of inter-relationships between cultured and other organisms, mediated by nutrients, energy, individual/group behaviour, competition, predation, external control.
Organism.	Context for function and integrity, interactions within habitat, ecosystems, intake of resources, output of growth, behaviour, reproduction; genetic and biochemical control.

be developed for aquaculture, motivate science and technology towards acquiring new skills and understanding, place these within the context of social attitudes and tolerance for the activity, in whatever form, allow access to various resources, and negotiate or accept the balance between the costs and benefits offered by the sector. This then sets the broader context in which the other sub-systems can be understood. From the 'bottom-up' direction, the various characteristics of the micro-systems determine the way in which they can be understood to exist and interact, the ways in which they can be located and managed, and the opportunities, consequences and desirability of intervention and management. Table 2.9 outlines some of the aims and potential advantages of adopting such an approach.

While the identification of the system assemblages and the depiction and understanding of their relationships with each other are important in developing choices, as outlined above, the characteristics of each system can also be identified more clearly, as outlined in Table 2.10, with examples for use in aquaculture and water resource management.

2.5.1   Practical applications

Systems analysis in the broader context now has a long-established precedence, from its original, mathematically-related applications, e.g. in process control (Forrester

Table 2.9 Aims and advantages of a systems perspective.

Aims, features	Issues/potential benefits
Definition and comprehension.	Description of system and sub-system structure helps to provide understanding of frameworks of context, cause and effect, implications of change at the system or subsystem level, effects of management or other intervention.
Elements and inter-relationships	Description and characterisation of intra-system elements and their inter-relationships provides the basis from which system behaviour can be understood, hence effects on other systems.
Functions and controls.	These determine the characteristics of inter-relationships, their significance, their potential for management, response to external perturbation, stability, etc.
Implications and impacts.	These are usually defined in the context of an internal (sub-) system or an external system, and provide the means to specify, quantify, the potential outcome of system change.
Context and explanation.	A more detailed description of a system, including its internal elements, provides more explanatory power, and/or allows hypotheses to be tested.
Prioritisation and choice.	An understanding of the complete systetms context, from the social/ethical to the molecular, allows areas of choice to be identified and their implications to be understood through the system chain.

1961), to its developments in more comprehensive social/ecological areas such as ecological systems (Odum 1971) and environmental management (Bennet & Chorley 1978). In contrast, the use of comprehensive systems perspectives in aquaculture and resource management has been rather limited (Muir 1992), though the descriptions of farming systems (Duckham & Masefield 1970; Dent & Anderson 1971), the ecology of agricultural production systems (Bayliss-Smith 1982), and descriptions of managed aquatic ecosystems (Michael 1987) have received more specific focus. In aquaculture, production systems have been described at the level of basic intensiveness of management and output, use of resources and levels water exchange (Hepher 1985; Muir 1992), in terms of closed cycle (recycle) production systems (Muir 1982; Timmons & Losordo 1994). There has also been a considerable body of work on integrated aquaculture systems (Pullin & Shehadeh 1980, Edwards *et al.* 1988), as well as systems approaches to components in aquaculture, such as cage culture/environment systems (Beveridge 1984), water/sediment systems (Moriarty & Pullin 1987) and pond eco-systems (Piedrahita *et al.* 1984), As Table 2.11 shows, the various types of aquaculture production can be described in simple systems terms.

Ecological systems analogies have been developed to some extent to describe aquaculture systems, as outlined in Table 2.12, and these have also been extended to link ecological and economic systems perspectives (Folke & Kautsky 1992). The modelling of various aquaculture-based interactions has also been widely considered (Cuenco 1989), and pond systems have received attention in view of their modified

Table 2.10 System characteristics to be considered.

Characteristics	Implications
Scale, extent, action.	Size and description of system, comprehensiveness, need for internal detail; e.g. hydrological regimes, consumer markets, occupied habitats, species assemblages, corporate entities, production units, stock groups.
Interlinkages, overlap.	Nature of linkages, extent to which one system description overlaps another; e.g. disease management, production planning, corporate dynamics, sectoral growth; identification of quantitative features (e.g. market sizes and consumer attributes, population, speciation and distribution of threatened species, water resource.
Quantitative/qualitative.	Availability, quality and variability, energy consumption and thermodynamic quality, stock size, type and market attributes, food and waste masses and compositions.
Dynamics/stability/ resilience.	Description of system characteristics, variability exhibited, response to change; e.g. effects of social/political change on markets, attitudes to environment, water management on catchment, stress/disease on stock, disease loss on corporate survival; nutrient change on receiving waters.
Determinism/predictability.	Extent to which system behaviour can be predicted, in general or specific terms, following external change; will depend on openness of system and whether change is local or global; e.g. market demand, price and business viability; disease, food wastage and environmental change.
Information/applicability.	Extent to which such descriptions can be developed on the basis of accessible information, and offer outputs which can be applied in a practical manner, e.g. through simulation, modelling; identifying critical and non-critical elements, e.g. socio-political change, market behaviour, economic growth, corporate planning, environmental capacity modelling, stock growth.

ecosystem properties, particularly under the United States Agency for International Development/Collaborative Research Support Programme (USAID/CRSP) (S. Nath, pers. comm.) using increasingly sophisticated structures to link productivity to external conditions, inputs and management strategies. Socio-economic impact assessments (e.g. UNDP/NMDC/FAO 1987; Garvey & Bennet 1991) may incorporate some element of systems understanding, with quantitative and qualitative elements. Resource use analyses, including energy analyses, studies of comparative efficiency of food production and protein production, may also provide useful perspectives towards systems assessment (see e.g. CGIAR 1987; Folke 1988). Economic system models have also been constructed concerning issues such as feeding and harvesting strategies in intensive aquaculture (Bjorndahl 1990), and more recently, models have been developed for farmer choice and decision-making in rural agriculture (Scoones & Thompson 1994).

Computer-based management information and decision support systems methodologies are also becoming more common in areas such as aquaculture (Turban 1988; Muir & Bostock 1994; Spratt 1994). Finally, as outlined in Table 2.13, a range of

Table 2.11   Systems descriptions of selected forms of aquaculture production.

Systems	Key external inputs	Internal mechanisms	Variability and control	Characteristics
Extensive pond.	Solar energy, rainfall/tidal cycle; stock rate, harvest timing, predation.	Pond ecosystem, air/water/benthos interface processes, stock metabolism.	Site-dependent; can be high external variability.	Moderate nutrient efficiency; productivity, output, usually high local sustainability.
Semi-intensive pond.	Water exchange, solar energy, stock and feed rate, predation.	Stock metabolism, pond ecosystem, air/water/benthos interface processes.	Less site-dependent, can be high internal variability, though external effects also important.	Moderate/good efficiency; good balance between resource use and delivery of social benefits.
Intensive pond/tank.	Stock rate, feed rate and composition, water exchange, oxygen inputs.	Stock metabolism, mixing and dispersal processes.	Slight site dependence, more external control, less mediation by ecosystem processes.	Poorer nutrient efficiency, high productivity, high external dependence.
Intensive marine cage.	Stock rate, feed rate and composition, tidal water exchange, incident energy.	Stock metabolism, mixing and dispersal processes, structural response.	Some site dependence, significant external control, variable ecosystem response.	Poor resource efficiency, high yield/financial efficiency, external dependence, varying local sustainability.
Mussel raft.	Tidal water exchange, incident energy and nutrient levels, stock rate, predation.	Stock metabolism, mixing and dispersal processes, structural response.	Site-dependent; stock and predation management may be critical.	High resource efficiency, moderate financial return, varying local sustainability.

systems-based perspectives can be described in terms of their potential relevance and use for the aquaculture sector.

2.5.2 Systems modelling

While systems descriptions may be offered for aquaculture and for its relationships with areas such as water resource management, there is then the question of what they can tell the observer. There has been a widespread interest in applying modelling approaches to systems of various types (Bennet & Chorley 1978), either functional/deterministic models where the working could be defined and outputs predicted from inputs, or 'black-box' stochastic models, which accept indeterminacy but attempt to describe output in at least generalised or probabilistic terms.

For environmental issues, models are commonly developed using the criteria and

Table 2.12 Intensive monospecies aquaculture as a stressed ecosystem.

Factor/characteristic	Features
Energy.	High auxiliary energy input, short residence time.
Nutrient/production.	Increase in exported/unused primary production, increased nutrient turnover, nutrient loss, reduced resource efficiency.
Species characteristics.	Higher proportion of growth-strategy species; increased parasitism, disease, other negative interactions.
Ecological characteristics.	Horizontal, one way transport; decreased vertical cycling, shortened food chains.
System characteristics.	Few simple, rapid, open/leaky cycles; networks with low average mutual information; simple system structures with few levels, low complexity, diversity, system efficiency; throughput based as low internal cycling.

Developed from Folke & Kautsky (1992)

Table 2.13 Examples of systems vs singular objectives relevant to aquaculture.

Subject	Singular	Systems
Politics.	Issue-based, personalities, 'arguments of the day'.	Strategic, thematic, developed from principles.
Economics.	Business performance, economic output, employment.	Well-being assessment, energy and life cycle analysis.
Development.	Personal/small group needs, project-based goals.	Community benefits, programme-based goals.
Planning.	Sector or project-based, polar responses; issue-defined.	Strategic, integrated, interrelation, synergy, net total gain.
Environment.	Issue-based, single species orientated, preservationist.	Resource, habitat/community based, conservationist.
Aquaculture.	Site and project-based, size, profitability, ratios.	Long-term performance, synergy, community benefit.
Sites.	Access, exposure, temperature.	Comparative advantage, sustainability.
Systems.	Initial cost, capacity.	Reliability, long-term effectiveness, adaptability.
Seedstock.	Size, survival to sale.	Quality, ongrowing performance.
Feeds.	Growth, FCR, digestibility.	Cost/gain, net environment cost.
Disease.	Diagnosis and cure, cost.	Predisposition, environment, management strategy.

conventions of systems theory, as applied to geographical and ecological systems. These range from the more simple conceptual structures to fully assembled compartment models that attempt to define stock, flows, transfer functions, etc. Such models have been developed for a range of aquatic resource issues, ranging from watershed models through lake process models to community and ecosystem approaches. These models have a far more clearly dimensional characteristic, and so can be used to handle both spatial and temporal effects. However, they are complex in their organisation and operation, and this complexity, while offering useful descriptive sense and forcing recognition of the various linkages, poses significant practice problems.

Thus, while such classical systems models may reveal interesting insights into systems behaviour, critical linkages and constraints, they may have limited practical application. According to Costanza *et al.* (1993), models must be recognised as only crude simplifications of reality, used to help to describe and understand elements of complex systems. While invaluable in helping to understand the relationships within and between systems, they can also be misused, and should be used to inform rather than legitimise management or policy decision. For models in ecological/economic systems, in which aquaculture could be included, three criteria were identified by which models could be evaluated:

- realism (simulating systems behaviour in a qualitatively realistic way),
- precision (quantitatively precise simulation), and
- generality (representing a broad range of system behaviours with the same model).

No single model can maximise all these goals, and so the choice of model and approach depends on the objective. A criterion for descriptive robustness can be applied, where a robust theorem will produce similar results using different models with different assumptions.

For strategic issues such as aquaculture and water resource management it may often be more useful to consider a matrix defined by geographical or spatially-based systems (Meaden & Kapetsky 1991; Ross *et al.* 1993), which describes locational constraints with respect to the presence and availability of combinations of resources, the quality of these resources, and the effects on the economic production function of the activity concerned (Muir 1994).

Whilst descriptions such as these may be constrained in describing socio-cultural, climatological or ecological perturbations, they might provide a starting point for describing critical interlinkages, and for developing critical choice criteria which may not be evident from simpler models or locational depictions. Whilst all of these elements can contribute usefully towards a systems perspective on aquaculture and water resources management, and hence to the wider but very clearly linked issues of sustainability, there have been few if any attempts to draw out a comprehensive perspective on the sector. This is perhaps unsurprising, given the relatively recent development of many of the concepts and areas of understanding, and the obvious complexity of the task.

2.6 FUTURE PERSPECTIVES

To meet the demands of future generations, assuming similar *per caput* levels of fish consumption (i.e. without even satisfying the unmet needs of poorer societies and communities and hence meeting only part of the overall sustainability demand), aquaculture production would have to expand significantly (Ruckes 1994). In approximate terms, assuming similar or even reduced levels of fishery output, output might have to rise by some 50% above 1990 levels by the year 2000, and in the 25 years from now, depending on demographic out-turns, might easily be required to double.

In broad terms, the macro-level sustainability description as outlined early in Fig. 2.1 provides a context for analysis, and a depiction of the ethical/societal attitudes to choice suggests prevailing economic cultures to be characterised as somewhere between technocentric and precautionary. The position held will be subject to change and influence, and as in any system may not necessarily be stable. There is probably currently a shift towards precautionary views, but this may easily swing back in the face of major conflict or as a result of dramatic technological change. A systems perspective would suggest that this cultural position could be inconsistent or self-contradictory; societies may hold concern for certain issues but may not be prepared to accept the consequences of that concern, or may be content to protect some elements but not others. Information and education, in a broad sense, would be important in any attempt to create a rational and coherent *Weltanschanung* for sustainable development.

Whatever the prevailing attitude to sustainability, it is clear that natural resources are being and will have to be employed, and to varying degrees traded, for physical/economic capital, and in turn human and societal benefit. In broad terms, current sustainability aims would allow natural resource use, with as little use as possible of the non-renewable, unless compensated for by longer-term gains in tradeability between renewable sources and social or economic benefits. There could also be scope for some diversity in sustainability, in that activities which measure poorly on these criteria may be compensated by others which produce a surplus, either within the sector (for example by using aquaculture methods or understanding to improve or rehabilitate natural resources) or by applying some of the material or cultural gains derived towards compensatory aims.

Within such a context, aquaculture and its development will be by no means unwelcome. It depends intimately on good quality natural resources, and involves a range of skills associated with the management of environments and the support of aquatic life. There is therefore an implicit involvement in sound management of resources, and hence a long term commitment towards sustainability. A systems view of the relationship between aquaculture and water resources might suggest that the more it develops, the greater will be the need for sustainable practice. Nevertheless, greater aquaculture output will clearly require additional resources, and it will require planners, investors, developers and consumers to be involved in the issues of choice. Aquaculture is but one of a range of activities with which individuals, families, communities or societies might be concerned. It may justify the time and resources applied, and may deliver the benefits desired, but must do so in competition with other activities (Domagala & Jaworski 1986). In spite of the many conceptual and practical difficulties, there are great benefits in adopting a systems perspective for evaluating aquaculture *vis-à-vis* water resource management, and to involve social, political, economic and natural resource factors.

Aquaculture can, then, be identifiable as a distinctive sub-component within wider systems, and can be treated in terms of its system structures and properties, its input–output relationships, its relative values, and its dynamics of change. Similar concepts could be applied broadly to other uses for resources to ensure that aquaculture is given the appropriate priority. To do this more effectively, further and more rigorous, testable and robust system concepts would be needed to provide reliable policy indicators. It will also be necessary to develop effective allocation frameworks in which the relative costs and benefits, in the widest sense, can be judged. Table 2.14 outlines some of the directions in which aquaculture systems research might be developed.

A systems approach adopted at the production level should be able to focus more clearly on more closely controlled systems, or more clear-cut understanding of the relationships and tradeoffs between resource mixes and management skill, including concepts of systems which are lower-yielding on area-based terms, but may be more efficient transformers of inputs, with less external impact. There may be more opportunities for integration at the designed and incidental level, for which a systems understanding should be able to determine more clearly the routes to overall efficiency.

In production terms, following the application of macro-level perspectives to micro-level analyses, aquaculture systems may be modified and developed, so that various levels of sustainability objective can be achieved. In this sense, local sustainability, as defined by system boundaries at the local level, may be easier to determine, and in terms of resource and energy flow the development of improved management techniques may satisfy many of the shorter term societally defined aims. However, even such locally sustainable systems may be subject to larger scale external change, owing to major development/resource use factors. For the more intensive aquaculture systems in particular, global sustainability issues may then be more difficult to satisfy.

Finally, it may be hoped that more comprehensive and explanatory systems descriptions will gradually emerge, and that these can be applied more widely in promoting development. The use of rational linkages between scale levels, the

Table 2.14 Aquaculture system research directions.

Area of research	Key features	Implications
Practical systems		
Productivity and feeding selectivity trials.	Types and quantities of feeds consumed by stocks within natural and managed ecosystems, hence manipulation potential.	Improved management capability in extensive and semi-intensive systems; productivity, better resource use, wider species strategies.
Basic polyculture trials.	Effects of species combinations in terms of feeding, behaviour, management, productivity, economics.	Better practical management, stocking and harvesting strategies, species choice, efficiency.
Integrated aquaculture trials.	Combinations of inputs, production systems, efficiencies, co-management strategies, economics.	Wider range of aquaculture opportunities, better resource use, complementary economic gain.
Integrated rural development trials.	Local resource development, social organisation, production, employment, response, benefit.	Better understanding of context, choice of development, extent and distribution of benefit.
Practical modelling		
Pond modelling.	Input, output and process functions, deterministic or stochastic approaches, stability, control.	Better management understanding, identification of key inputs, control variables, better decisions.
Environmental capacity modelling.	Indicators, relationships, risk analysis, modelling approaches, impacts, responses.	Better use and management of ecosystems, sustainability, multiple objective development.
Social system analysis.	Characterisations, indicators, impacts, responses, mechanisms.	Understanding of underlying processes in rural development, potential approaches.
Strategic modelling		
Techno-economic modelling.	Performance analysis, profitability, comparative efficiency, response to key inputs, risk response.	Better decision making in research strategy, allocation choice, sectoral strategy, sustainability.
Economic decision modelling.	For the individual, firm and/or sector; nature of choice and decisions, competition, risk, strategy.	Understanding of economic behaviour; better strategy, sectoral development, management.
Strategic resource mapping.	Inventories, relationships, presentation, simulations of choice and outcome, GIS, etc.	Effective strategic context for choice, allocation, multiple-objective development.
Systems theory		
General approaches.	Assess/use emerging concepts into systems/modelling for sector; include moral/ethical concepts.	New ways of viewing complex problems; understanding of limits of determinism.
Economic concepts.	Development/use of macro- and micro-economic approaches, environmental economics, others.	Alternative viewpoints from neoclassical economics potential for sustainability objectives, development.
Ecosystem theory.	Understanding of key features and functions, evolutionary mechanisms, controls, ecosystem risks.	Understanding of ecosystem criteria, risks, value of indicators, objectives.

understanding of systems inter-relationships across social, ecological and economic boundaries, and the application and proof of more effective production methods, may all be expected to contribute towards a global level of sustainability for aquaculture.

REFERENCES

Bailey, C. & Skladany, M. (1990) Sociology and aquaculture: the political economy of aquaculture development in the Third World. *ICA Communicae, Auburn University International Center For Aquaculture*, **12**(1–2), 6–8.

Barg, U.C. (1992) Guidelines for the promotion of environmental management of coastal aquaculture development. *FAO Fisheries Technical Paper*, **328**,. FAO, Rome.

Bayliss-Smith, T.P. (1982) *The Ecology of Agricultural Systems.* Cambridge University Press, Cambridge, UK.

Bennett, R.J. & Chorley, R.J. (1978) *Environmental Systems: Philosophy, Analysis and Control.* Methuen and Co, London.

Beveridge, M.C.M. (1984) Cage and pen fish farming. Carrying capacity models and environmental impact. *FAO Fisheries Technical Paper*, **255**. FAO, Rome.

Beveridge, M.C.M., Ross, L.G. & Kelly, L.A. (1994) Aquaculture and biodiversity. *Ambio*, **23**(8), 497–502.

Bjørndahl, T. (1990) *The Economics of Salmon Aquaculture.* Blackwell Scientific Publications Ltd, Oxford.

Briggs, M.R.P. (1993) Water quality in sustainable shrimp farming. In: *The Third Academic and Research Conference for Southern Thailand Technical Colleges.* Krabi, Thailand, 23–25 July 1993.

CGIAR (Consultative Group on International Agricultural Research) (1987) CGIAR priorities and future strategies. TAC Secretariat, FAO, Rome.

Chambers, R. & Conway, G.R. (1992) Sustainable rural livelihoods: practical concepts for the 21st century. *Institute of Development Studies, Discussion Paper*, **296**. Institute of Development Studies, Falmer, Sussex.

Chua, T.E. (1993) Environmental management of coastal aquaculture development. In: *Environment and Aquaculture in Developing Countries* (ed. R.S.V. Pullin, H. Rosenthal & J.L. Maclean), pp. 199–212. *ICLARM Conference Proceedings*, **31**. ICLARM, Manila.

Costanza, R., Wainger, L., Folke, C. & Grogan-Maler, K. (1993) Modelling complex ecological-economic systems. *Bioscience*, **43**(8), 545–55.

Cuenco, M.L. (1989) Aquaculture systems modelling: an introduction with emphasis on warmwater aquaculture. *ICLARM Studies and Review*, **19**. ICLARM, Manila.

Dent, J.B. & Anderson, J.R. (eds) (1971) *Systems Analysis in Agricultural Management.* John Wiley and Sons, Sidney.

Domagala, M. & Jaworski, K. (1986) Guidelines for the calculation of the comparative costs of animal protein in developing countries. *FAO Fisheries Circular*, **802**, FIPP/C802. FAO, Rome.

Duckham, A.N. & Masefield G.B. (1970) *Farming Systems of the World.* Chatto and Windus, London.

Edwards, P., Pullin, R.S.V. & Gartner, J.A. (1988) Research and education for the development of integrated crop–livestock–fish farming systems in the tropics. *ICLARM Studies and Reviews*, **16**. ICLARM, Manila.

FAO (1992) Aquaculture production 1984–1990. *FAO Fisheries Circular*, **815** Revision 4, FIDI/C815 (Rev. 4). FAO, Rome.

FAO (1995) Aquaculture production 1984–1993. *FAO Fisheries Circular*, **815** Revision 7, FIDI/C815 (Rev. 7). FAO, Rome.

Folke, C. (1988) Energy economy of salmon aquaculture in the Baltic Sea. *Environmental Management*, **12**(4), 525–37.

Folke, C. & Kautsky, N. (1992) Aquaculture with its environment: prospects for sustainability. *Ocean and Coastal Management*, **17**, 5–24.

Forrester, J.W. (1961) *Industrial Dynamics.* Massachusetts Institute of Technology Press, Massachusetts.

Garvey, S. & Bennet, J. (1991) Aquaculture and rural development in the west of Ireland. *Proceedings of the EAS Aquaculture Conference*, Dublin.

GESAMP (1991) Reducing environmental impacts on coastal aquaculture. *Reports and Studies*, **47**. FAO, Rome.

Gow, D.D. (1992) Poverty and natural resources: principles for environmental management and sustainable development. *Environmental Impact Assessment Review*, **12**(1/2), 49–65.

Håkanson, L., Ervik, A., Mäkinen, T. & Moller, B. (1988) Basic concepts concerning the environmental effects of marine fish farms. *Nord 1988*, **90**. Nordic Council of Ministers, Copenhagen.

Hepher, B. (1985) Aquaculture intensification under land and water limitations. *Geojournal*, **10**, 253–9.

Holling, C.S., Schindler, D.W., Walker, B.W. & Roughgarden, J. (1994) Biodiversity in the functioning of ecosystems: and ecological primer and synthesis. In: *Biodiversity Loss: Economic and Ecological Issues* (ed. C. Perrings, C. Folke, K.G., Maler, B.-O. Jansson & C.S. Holling), Cambridge University Press, Cambridge.

Jacobs, P., Garner, J. & Munro, D.A. (1987) Sustainable and equitable development. In: *Conservation with Equity* (ed. P. Jacobs & D.A. Munro), *Proceedings of the Conference on Conservation and Development – Implementing the World Conservation Strategy*, Ottawa 31 May–5 June 1986. International Union for the Conservation of Nature, Cambridge, UK.

Jensen, A. (1993) Carrying capacity for nutrient salts in coastal waters. In: *Fish Farming Technology* (ed. H. Reinertsen, L.A. Dahle, L. Jorgensen and K. Tvinnereim), pp. 151–4. A.A. Balkema, Rotterdam.

Leach, G. (1976) *Energy and Food Production*. IPC Press, Guildford.

Lele, S.M. (1991) Sustainable development: a critical review. *World Development*, **19**(6), 607–21.

Meaden, G.J. & Kapetsky, J.M. (1991) Geographical information systems and remote sensing in inland fisheries and aquaculture. *FAO Fisheries Technical Paper*, **318**. FAO, Rome.

Michael, R.G. (1987) (ed.) Managed aquatic ecosystems. *Ecosystems of the World*, **29**. Elsevier, Amsterdam.

Moriarty, D.J.W. & Pullin, R.S.V. (eds) (1987) *Detritus and Microbial Ecology in Aquaculture. ICLARM Conference Proceedings*, **14**. ICLARM, Manila.

Muir, J.F. (1982) Recirculated water systems in aquaculture. In: *Recent Advances in Aquaculture* (ed. J.F. Muir & R.J. Roberts), pp. 357–447. Croom Helm, London.

Muir, J.F. (1992) Aquaculture and water resource management. In: *Proceedings of the Conference on Priorities for Water Resources Allocation and Management*, pp. 127–48. ODA Natural Resources and Engineering Advisers' Conference, Southampton, July 1992. Overseas Development Administration, London.

Muir, J.F. (1994) *Sustainable Systems of Land and Water Use in Scotland: Freshwater Resources*. Report commissioned by Scottish Natural Heritage and Scottish Office Agriculture and Fisheries Department, Edinburgh.

Muir, J.F. & Baird, D.J. (1991) *Environmental Management of Aquaculture Development. FAO Technical Report, TCP/CYP/9152*. FAO, Rome.

Muir, J.F. & Bostock, J.C. (1994) Measurement and control in aquaculture processes – new directions for the information revolution? In: *Proceedings, EAS Conference, 'Measures for Success' Instrumentation and Control in Aquaculture*, Bordeaux, 23–25 March 1994. *EAS Special Publication*, **21**. Oostende.

Odum, E.P. (1971) *Fundamentals of Ecology*, 3rd edn. Saunders, Philadelphia.

Pearce, D. (1991) *Blueprint 2. Measuring Sustainable Development*. Earthscan, London.

Peet, J. (1992) *Energy and the Ecological Economics of Sustainability*. Island Press, Washington DC.

Piedrahita, R.H., Brune, D.E., Tchobanoglous, G. & Orlob, G.T. (1984) A general model of the aquaculture pond ecosystem. *Journal of the World Mariculture Society*, **15**, 355–66.

Pimentel, D. & Pimentel, M. (1979) *Food, Energy and Society*. Edward Arnold, London.

Pullin, R.S.V. & Shehadeh, Z.H. (eds) (1980) Integrated agriculture–aquaculture farming systems. *ICLARM Conference Proceedings*, **4**. ICLARM, Manila.

Pullin, R.S.V., Rosenthal, H. & Maclean, J.L. (eds) (1993) *Environment and Aquaculture in Developing Countries*. ICLARM Conference Proceedings, **31**. ICLARM, Manila.

Reid, D. (1993) *From Nature Conservation to National Sustainable Development Strategies: A Review of the Literature of Nature Conservation, Sustainable Development and Sustainability 1966–93*. Centre for Human Ecology, University of Edinburgh, Edinburgh.

Ross, L.G., Mendoza, A.Q.M. & Beveridge, M.C.M. (1993) The application of Geographical Information Systems to site selection for coastal aquaculture: an example based on salmonid cage culture. *Aquaculture*, **112**, 165–78.

Ruckes, E. (1994) Can aquaculture fill the market gap? *FAO Aquaculture Newsletter*, April 1994, No 6, 17–18.

Scoones, I. & Thompson, J. (eds) (1994), *Beyond Farmer First. Rural People's Knowledge, Agricultural Research and Extension Practice*. Intermediate Technology Publications, London.

Shopley, J.B. & Fuggle, R.F. (1984) A comprehensive review of current EIA methods and techniques. *Journal of Environmental Management*, **18**, 25–47.

SNH (1993) *Sustainable Development and the Natural Heritage*. Scottish National Heritage, Edinburgh.

Spratt, M. (1994) Modelling aquaculture production with spreadsheets. *Proceedings, EAS Conference, 'Measures for Success' Instrumentation and Control in Aquaculture*, Bordeaux, 23–25 March 1994. *EAS Special Publication*, **21**. Oostende.

Tacon, A.G.J. (1994) Dependence of intensive aquaculture systems on fishmeal and other fishery resources. *FAO Aquaculture Newsletter*, April 1994, No 6, 10–16. FAO, Rome.

Timmons, M.B. & Losordo, T.M. (eds) (1994) *Aquaculture Water Reuse Systems: Engineering Design and Management. Developments in Aquaculture and Fisheries Science*, **27**. Elsevier, Amsterdam.

Turban, E. (1988) *Decision Support and Expert Systems*. Macmillan, New York.

UNDP (United Nations Development Programme) (1994) *Human Development Report*. Oxford University Press, New York, Oxford.

UNDP/NMDC/FAO (1987) *Thematic Evaluation of Aquaculture*. FAO, Rome.

WCED (World Commission for Environment and Development) (1987) *Our Common Future*. Oxford University Press, New York, Oxford.

World Bank (1992) *World Development Report 1992. Development and the Environment*. Oxford University Press, New York, Oxford.

Chapter 3
Waste Production from Aquaculture

Asbjørn Bergheim   *Rogaland Research, Stavanger, Norway*
& Torbjørn Åsgård   *Akvaforsk, Sunndalsøra, Norway*

3.1 INTRODUCTION

Wastes from aquaculture include all materials used in the process which are not removed from the system during harvesting. The quantity of the total waste produced, which leaves the system to load the environment, is closely correlated to the culture system used.

The environmental impacts of different aquaculture systems are described by Pullin (1989) and Braaten & Hektoen (1991). Serious pollution problems are associated with intensive systems, typically monoculture systems reliant on the input of artificial feed. Generally, intensive systems include most temperate zone finfish farms and an increasing number of shrimp pond farms in tropical and sub-tropical regions.

Semi-intensive systems dominate the tropical production of both freshwater finfish and coastal and freshwater shrimp. The main available inputs are natural feed sources suitable as aquatic feeds, including fresh and composted vegetation, cereal brans and oil cakes, and fertiliser (including animal manure, human sewage and chemical fertiliser). The waste output from these systems is much less than that from highly-intensive feedlot-based systems.

Differences in both the quality and quantity of the waste components are dependent on the aquaculture system and species cultured (Pillay 1992). Most of the available information on waste production comes from intensive systems, particularly for the production of salmonids in hatcheries and net pen farms. The environmental impacts from tropical aquaculture have recently been described by Pullin *et al.* (1993).

The principal wastes from aquaculture are uneaten feed, excreta, chemicals and therapeutics (Beveridge *et al.* 1991). The term 'waste' can also refer to dead and moribund fish, and even escaped fish and pathogens. This paper presents a brief review of published material describing wastes derived from feeds, fertilisers and residues from biocides and chemicals. The available knowledge is especially related to intensive rearing systems for salmonids, but the information about waste production from other systems, such as tropical semi-intensive facilities, is also included.

3.2   THE ORIGIN OF WASTES

3.2.1   Feed-derived wastes

The idea of aquaculture is to convert in a profitable way energy and nutrients in feed or fertiliser into energy and nutrients in an edible product, with minimum loss in the conversion.

The waste from intensive aquaculture plants is predominantly from feed, and includes uneaten feed (feed waste), undigested feed residues and excretion products (Cripps 1993). Factors connected to feedstuff quality, diet formulation, feed production technology and feeding practice greatly influence the total waste loading.

The relationship between energy fed and energy expended is shown in Fig. 3.1. Energy can be lost in uneaten feed (feed waste), in the undigested fraction of uneaten feed or in the excretion products from protein used as a source of energy. Feed waste will increase rapidly when energy fed exceeds maximum energy intake and will become the main fraction of organic loss. The organic loss can cause oxygen consumption in the water when it is metabolised by other organisms. The energy spent in basal metabolism, in activity or in catabolic processes during digestion is lost as heat. This loss of energy is related to oxygen consumed and carbon dioxide and water produced by the fish. All energy from the feed is lost when the feed intake is below the maintenance level. When energy intake is above the maintenance level, some energy is used for growth and deposited as organic matter in protein, lipid or glycogen. The relative proportion of energy retained in body mass will increase as the feed intake rises from maintenance level to maximum intake.

For the nutrients nitrogen (N) and phosphorus (P) the picture is somewhat similar. The nutrients fed can be lost in feed waste or in faecal waste without being absorbed.

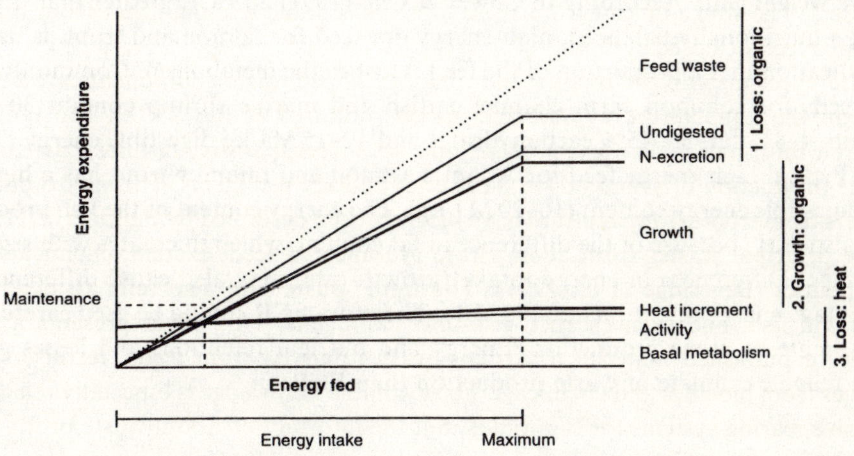

Fig. 3.1   Energy expenditure in fish at different feeding and intake levels; (1) lost in organic matter, (2) retained in growth, (3) lost as heat.

The absorbed nutrients can be spent on maintenance or retained in body mass increase. Endogenous losses of N and P are relatively small. If the amount of nutrients absorbed exceeds the requirements for maintenance and growth, the excess will be excreted. Since nutrient requirements for maintenance and growth are different, a diet designed to give balanced nutrition and minimum loss at one feed intake level will not be balanced at another intake level, and the losses will be different.

Feed type

This section deals mainly with waste production originating from nutritionally complete feed types used as the only nutrition source (except for minerals) in intensive aquaculture systems. Compounded feed can, however, also be given as a supplement to fertilisers in many types of pond culture. The addition of compounded feed in fertilised ponds for channel catfish and common carp can increase the harvested yield up to ten times that from unfed ponds (Lovell 1989).

To improve feed utilisation and reduce local pollution, the use of moist diets with poor consistency has been gradually replaced with nutritionally-improved semi-moist or dry feeds. Watanabe (1991) reported a greatly reduced feed conversion ratio (FCR) and effluent loadings of nutrients (N, P) from both yellowtail and red sea bream cultures, when raw minced fish was replaced with semi-moist or dry feed. Diet replacement in Norwegian salmon culture produced similar results, with an FCR reduced from 3.5 in 1975 to 1.1 in 1993. The reduction in the prevalent feed-derived pollution problems in Danish rivers downstream from trout pond farms was mainly solved by the introduction of improved feed quality (Kiaerskou 1991).

The utilisation of dry feed, often expressed as FCR, has been experimentally determined to be within the range of less than 1.0 to greater than 2.0 kg dry feed per kg live weight gain. According to Cowey & Cho (1991) an FCR greater than 1.0–1.2 using a nutritionally balanced, high-energy dry feed for salmon and trout, is usually an indication that a proportion of the feed is lost to the metabolism. Commonly used dry feeds for common carp, channel catfish and marine shrimp contain 30–40% protein, 4–8% fat, 30–45% carbohydrates and 12–15 MJ/kg digestible energy (Table 3.1). Typical high energy feed for Atlantic salmon and rainbow trout has a high fat and digestible energy content (16–20 MJ/kg). The energy content of the fish produced may also vary, because of the difference in fat content, which fluctuates with size and life stage. Differences in energy intake (growth rate) may also cause differences in percentage energy retention (see Fig. 3.1). Therefore FCR should be used carefully as an estimate of waste production. Energy and nutrient retentions and losses give a more reliable estimate of waste production than FCR.

Feed waste and ingested feed

The total amount of feed given is often assumed to have been eaten. It is, however,

Table 3.1 Reported dry feed diets and feed utilization values (FCR) for different aquaculture organisms.

Fish species	Feed description	Feed composition, % of wet weight					FCR*	Reference
		Dry matter	Protein	Fat	Carbohydrates (fibres)	Phosphorus (available)		
Common carp	Commercial	90–92	35	8	34	1.7–2.2	1.3–1.4	Watanabe (1991)
Common carp	Experimental (plant protein)	—	6.5 / 30		(6.4)	(0.7)	1.9–2.3	Akiyama (1991)
Channel catfish	Commercial	—	31–35	4.3–5	(3.3–5.3)	(0.4–0.5)	1.6	Akiyama (1991)
Marine shrimp	'Pelleted feed'	90	45	6.1	23 (3.1)	1.3	2.0	Briggs & Funge-Smith (1994)
Salmonids								
Rainbow trout	Commercial	—	47	24	11	1.0	1.0	Alsted (1991)
Rainbow trout	Commercial	91	42	21	23	—	1.1–1.5	Alanärä (1992)
Atlantic salmon	Experimental	92	46	22	17 (1.1)	1.2	1.26	Johnsen, Hillestad &
		92	31	30		1.1	0.96	Austreng (1993)

* FCR = feed conversion ratio, kg feed/kg weight gain (WW based)
—: no figures

difficult to determine the quantity of uneaten feed during normal commercial farming conditions. If the fish are fed to satiation, it is difficult to avoid some feed wastage. There are, however, systems developed for the detection and collection of feed waste in certain culture enclosures.

This feed waste can contribute considerably to the waste loadings from a fish farm (Åsgård *et al.* 1991). The stability of the feed in water is an important factor determining nutrient leakage to the environment. Even dry feed pellets can disintegrate within a few minutes in water, while other, apparently similar, types of pellet maintain their structure for several days. Water stability is vital for the efficiency of installed feed waste collector systems. In addition, feed particles not eaten immediately can be consumed later by the culture organisms, or other higher organisms, if the consistency of the feed pellet can be maintained for some time. Good feeding practice and efficient feed waste detection and/or collection systems (Blyth & Purser 1993; Helland & Grisdale-Helland 1993) reduce losses to a minimum.

The main waste components from eaten feed are undigested feed (fecal waste) and the excretion products from the catabolised proteins, such as total ammonia nitrogen (TAN) and urea. Also excreted minerals, especially phosphorus, are present. An example of the relative contribution of the different sources, when using a high energy diet, is presented in Table 3.2 (Johnsen *et al.* 1993). Based on ingested feed, the production of 1 kg biomass yields an estimated fecal waste quantity of 162 g organic matter, comprising 50 g protein, 31 g lipids and 81 g carbohydrates. The waste production of nutrient salts was 30 g total nitrogen (TN), comprising 8 g protein N + 22 g total ammonia N (TAN) and 7 g total phosphorus (TP). In this example, about 51% and 64% of the supplied N and P respectively were considered lost. Under actual farming conditions, the waste production can be considerably higher. In 1990 approximately 70% of N, P and organic matter (energy) from commercial feeds in Norwegian salmon culture ultimately caused a load on the environment (Austreng & Åsgård 1991). This may still be a common situation in many areas where conventional feeds are used.

Table 3.2 Estimated mass balance for the production of 1 kg of Atlantic salmon (size 1.2 kg) calculated from Johnsen *et al.* (1993).

Nutrients in feed	Ingested nutrients (g)	Retained in fish (g)	Feces loss (g)	Excretion (g)
Protein N	59	29	8	22
Lipid	286	138[a]	31	[b]
Carbohydrate	152	—	81	[b]
Total P	11	4	6[c]	1
Organic matter	805	319	162	[d]

[a] Estimate – based on a growth trial with simiilar fish and diet (T. Åsgård, unpublished data)
[b] Spent energy equivalent to 117 g lipid and 71 g carbohydrate
[c] Estimate – based on a digestibility trial with similar fish and diet (T. Åsgård, unpublished data)
[d] Equivalent to 324 g organic matter spent as energy and in N excretion products

In practice, the digestibility of the dietary nutrients varies for several reasons. The diet is usually composed of several feedstuffs. The digestibility of the nutrients in a single feedstuff may vary with fish age and species. Further, the digestibility may be influenced by factors such as the source of the feedstuff, date of capture of the feed species, quality of the raw material, process technology, inclusion rate in the diet and interaction with other feed ingredients. The use of binders and the recent development in feed extrusion and fat inclusion technology have greatly affected nutrient availability. The reported digestibility range of commonly used protein sources, of both plant and animal origin, by carp, channel catfish, rainbow trout and tilapia is 75–100% (Ash 1985; Åsgård 1988; Steffens 1989). Lipids with low melting point have a high digestibility (Austreng *et al.* 1980). Grass carp can digest 94–100% of the fat from fish meal and soya (Steffens 1989). The digestibility of low molecular weight carbohydrates is generally high in both salmonids and carp, while the digestion of raw polysaccharides is limited in salmonids such as rainbow trout (Cowey & Walton 1989). Carbohydrates in some processed products are, however, well utilised at moderate levels in diets (Arnesen 1993). Crude fibre is almost indigestible to all fish.

Beveridge *et al.* (1991) estimated the fecal production from different aquacultural organisms (Fig. 3.2), based on the digestibility of the principal constituents in typical commercial diets. Faecal dry matter production was calculated to be 26–27% of the ingested feed of salmonids, carp and shrimps,and 41% for catfish. The salmonid feed used had a moderate energy level (19% fat, 45% protein, 27% carbohydrate) and a lower overall digestibility than high energy feed (Table 3.2).

Several methods are used for determining digestibility (Austreng 1978; Cho & Slinger 1978; Choubert *et al.* 1979).

The bioavailability of dietary P is strongly influenced by chemical form, fish stock and other factors (Lall 1991). Soluble, inorganic P (as mono-calcium P, sodium P) is highly available to salmonids, carp and catfish (> 90% bioavailability). Plant P,

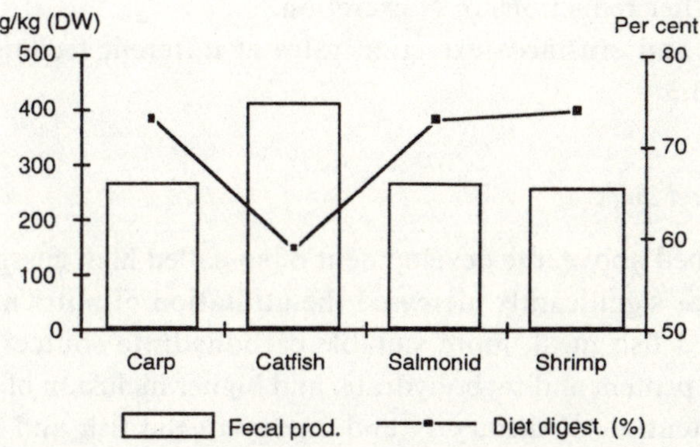

Fig. 3.2 Calculated fecal production (g DM/kg DM ingested feed) based on estimates of digestibility of typical diets. Data from Beveridge *et al.* (1991). The whole diet digestibility used: 73% in carp and salmonid, 59% in catfish and 74% in shrimp.

phytate P, is unavailable to most finfish. Lall (1991) reported a requirement of 0.4–0.8% bioavailable P in the feed for several aquaculture fish species. The dietary P requirement, expressed as a percentage of the total ration, will increase if the food conversion ratio (FCR) is improved, such as by using high energy diets. The digestibility of P in fish meal, the main P source in some diets, is often only 40–60%. The TP content in commercial diets is usually 1.0–2.2% (Table 3.1). Therefore, 50–80% of the supplied P is released to the recipient, because of either too low availability, or supply in excess of requirements.

Nitrogen excretion

Fish excrete a number of nitrogenous compounds, including ammonia, urea, trimethylamine, creatine and creatinine (see review by Handy & Poxton 1993). The main excretory product in teleost fish is ammonia (TAN), which is formed in the liver and excreted across the gills (Ramnarine *et al.* 1987). In farmed salmonids TAN and urea account for 80–90% and 10–20% of the total N excretion (fecal N not included), respectively (Fivelstad *et al.* 1990; Forsberg in press).

The daily excretion rate of N is directly related to the N consumed (Kaushik 1980; Forsberg in press). The N excretion from salmonids and carp fed to satiation is 5–15 times the excretion during starvation conditions (endogenous excretion). Feed composition is also important. Johnson *et al.* (1993) found that N excretion decreased from 35 to 22 g per kg of salmon produced, when the energy content of the feed was increased (fat increased 22–30%, protein reduced 44–38%). Simultaneously the undigested fecal N was reduced from 11 to 8 g per kg fish produced. If there is surplus protein or the amino acid profile of the feed does not correspond to the requirements of the fish, the excess becomes deaminated, and N is excreted. Since the protein : energy ratio in maintenance metabolism is lower than in growth metabolism, the optimal feed composition changes with feed intake (see Fig. 3.1). This indicates a potential for further reductions in N excretion.

Reported fish and crustacea excretion rates at different feeding rations are presented in Table 3.3.

Recently developed diets

As briefly described above, the development of so-called high energy diets, especially for salmonids, has significantly increased the utilisation of nutrients. The use of low temperature dried fish meal, more suitable carbohydrate sources, pellet extrusion, reduced levels of protein and carbohydrate, and higher inclusion of fat in the diet, has improved the retention of dietary N and energy in the fish and therefore reduced waste production (Alsted & Jokumsen 1989; Johnsen & Wandsvik 1991; Johnsen *et al.* 1993; Hillestad & Johnsen 1994). Using these high energy diets, a N retention of approximately 50% and energy retention of 50–55% was achieved during

Table 3.3 Excretion rates of nitrogenous compounds by different aquaculture species. Units: mg N/kg/h.

TAN = NH_3–N + NH_4–N   % of ingested = (TAN + event. Urea-N/total ingested N) × 100%

Group of species	TAN	Urea-N	% of ingested	Size, temperature, feed rate	Reference
Carps					
C. carpio L.	2.2–24.2	—	19.35	350 g, 16–18°C, 0–100%	Kaushik (1980)
'Carp'	3.0–3.6	—	(starving)	78–370 g, 20–27°C, 0%	Ogino et al. 1973
Tilapia					
Oreochromis mossambicus	0.83	0.53	(starving)	—, 13°C, 0%	Sayer & Davenport (1987)
O. niloticus	1.7–9.4	—	—	(fed fish)	cited in Beveridge & Phillips (1993)
Salmonids					
Rainbow trout	2.7–37.3	0.8–3.0	44–52	130 g, 15–18°C, 0–100%	Kaushik (1980)
Atlantic salmon	0.7–5.0	0.2–0.8	25–32	2 kg, 8.5°C, 0–100%	Forsberg (in prep.)
Atlantic salmon	2.8–6.0	0.4–0.9	25–38	2–4.2 kg, 4–12°C, 100%	Fivelstad et al., 1990
Anguilla anguilla	5–50	0–27	—	1–3 g, 25°C, 0–100%	Knights (1989)
Marine fish					
Gilthead seabream	11–43	0*	30	3–90 g, 21–24°C, 0–100%	Porter et al. 1987
Seabass	18–29	—	—	3–90 g, 24°C, 100%	Guérin-Ancey (1976)
Atlantic cod	6–41	—	33	200 g, 14°C, 20–100%	Ramnarine et al. 1987
Atlantic halibut	1.4–6.9	—	—	454–2334 g, 10°C, 0–100%	Davenport et al. 1990
Crustacea					
Penaeus monodon	12–39	—	—	1.6–27 g, 28°C, 100%	Wickins (1985)
Penaeus spp.	14–90	—	—	—	cited in Phillips et al., 1993

— no figures
* no detectable urea-N

experiments simulating practical farming conditions. Energy and N retention levels of even greater than 60% have been achieved with this type of diet in small fish (B. Grisdale-Helland & S.J. Helland, pers. comm.). This retention efficiency is double that regarded as usual for fish (Jobling 1993).

Even low-ash fish meal contains approximately 2% P. Fish meal must be partially replaced by other low P protein sources if the P content of a complete feed is to be substantially reduced. In an experiment with adult salmon, a 60% replacement of herring meal with soybean meal reduced the P discharge by approximately 50% (Bergheim & Sveier 1993). Growth rate and feed utilisation were still maintained. Meyer-Burgdorff (1992) achieved a similar reduction in P excretion from rainbow trout (*c*. 50%) by the partial substitution of fish meal by animal by-products (feather meal, meat meal). Addition of phytase has proved to be an effective means of increasing the P digestibility of both rainbow trout and carp feed containing soya as the main protein source (Meyer-Burgdorff 1992).

3.2.2 Pond fertilisation

The main sources of fertilisers used in semi-intensive aquaculture systems are live-stock manure, human sewage and chemical fertiliser. Animal manure is considered to be better than inorganic fertiliser at promoting the growth of planktonic and benthic food organisms in fresh- and brackish-water ponds (Pillay 1992). The use of manure as the main nutrient input is the traditional practice in Chinese pond farms stocked with a polyculture of carp. The system has been applied in other regions with different fish species and types of manure.

Livestock manures have a low metabolic energy and protein content compared with conventional feeds, and are considered to be indirect food for fish (Wohlfarth & Schroeder 1979). Little is known of the processes by which manure is converted into food (Wohlfarth & Hulata 1987), and the possible conversion pathways, including heterotrophic and autotrophic utilisation and direct consumption, have not been adequately quantified. The main pathways of manure breakdown and nutrient conversion in a typical polyculture of carp are illustrated in Fig. 3.3.

Both animal and human wastes have a great nutrient and degradable organic matter concentration (Table 3.4). The recommended quantity of manure applied to a balanced pond system should maintain stable concentrations of nutrients and phyto-plankton in the pond water, e.g. 19–39 mg/l of phytoplankton during the growing season where manure is applied at a rate of only 4–5 g manure $DM/m^3/day$ (Yang *et al*. 1992). Cereal products with different N concentrations, such as cottonseed meal and rice bran, have been used as fertilisers for the production of striped bass and mullet (Ludwig & Tackett 1991; Bishara 1979).

Indirect feeds are expected to be far less efficient than direct conventional feeding, since approximately 90% of the energy and nutrients become unavailable at every

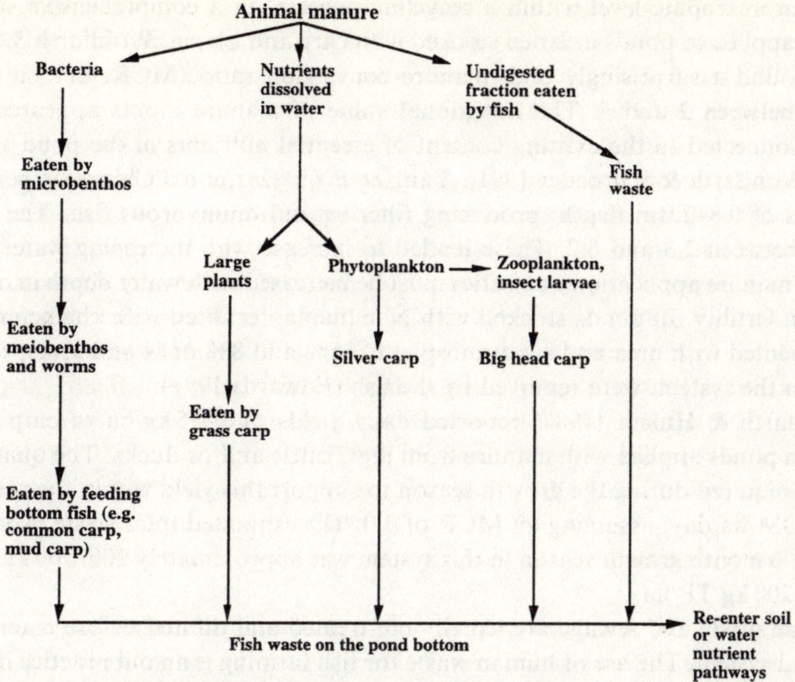

Fig. 3.3 Schematic breakdown of animal manure and nutrient pathways in ponds stocked with Chinese and common carp. Re-drawn after Delmendo (1980).

Table 3.4 Contents of organic matter and nutrients in commonly used organic fertilizer in pond fish farming. Source: animal and human waste calculated from Piedrahita & & Tchobanoglous (1987), rice bran and cottonseed meal (from Ludwig & Tackett (1991).

Organic fertilizer	Dry matter, %	Composition, g/kg wet weight				
		CBOD(5)	COD	TN	TP	TK
Fecal waste						
Dairy cattle	13	21	112	5.0	0.9	3.3
Swine	9	31	88	6.9	2.3	4.5
Poultry	25	47	160	9.5	3.8	3.8
Human	8	23	45	9.1	1.8	1.8
Rice bran	88	–	–	20	28	16
Cottonseed meal	91	–	–	70	14	20

CBOD (5): carbonaceous 5-day biochemical oxygen demand
COD: chemical oxygen demand

transition in trophic level within a recycling process. In a comprehensive study of manure applied to ponds in Israel stocked with carp and tilapia, Wohlfarth & Hulata (1987) found a surprisingly high manure conversion ratio (MCR, dry matter) of usually between 2 and 4. The nutritional value of manure inputs appeared to be closely connected to the existing content of essential nutrients in the pond soil and water (Wohlfarth & Schroeder 1991). Yang *et al.* (1992) quoted Chinese experiments in ponds of 0.8–2.0 m depth, producing filtering and omnivorous fish. The MCRs ranged between 2.3 and 6.2. These tended to increase with increasing water depth, because manure application quantities must be increased with water depth in order to maintain fertility. In ponds stocked with Nile tilapia, fertilised with chicken manure supplemented with urea and superphosphate, 15% and 8% of N and P respectively, added to the system, were removed by the fish (Edwards 1993).

Wohlfarth & Hulata (1987) reported daily yields of 7–35 kg/ha of carp and/or tilapia in ponds applied with manure from pigs, cattle and/or ducks. The quantity of manure required during the growth season to support this yield was in the range 20–100 kg DM/ha/day, assuming an MCR of 3.0. The estimated total waste production during a 6 month growth season in this system was approximately 200–700 kg TN/ha and 70–200 kg TP/ha.

Human waste and sewage are usually pre-treated and diluted before entering the fish pond system. The use of human waste for fish farming is an old practice in Asian countries, such as China, and is not basically different from the application of animal manure (Pillay 1992). Reported yields of carp and tilapia in sewage-fed ponds are 3–10 t/ha/year (Piedrahita & Tchobanoglous 1987) and are considerably lower in ponds used primarily for waste water treatment. The presence of pathogenic organisms in faeces is of great concern. Sewage should never be used for aquaculture without some form of pre-treatment such as settlement or oxidation to remove the pathogens (Buras 1993).

Commercial inorganic fertilisers are used infrequently, but to an increasing degree, in aquaculture in developing countries (Edwards 1993). In North America, compound fertilisers (NPK) seem to be favoured, while P fertiliser seem to be preferred among Western European aquaculturists (Pillay 1992). Addition of N fertiliser to ponds fertilised with P and K can increase the production of planktivorous species such as tilapia, yielding economic benefits (Boyd 1976).

The dosage of fertiliser has usually to be adjusted according to factors including the nutrient status of pond water and soil and grazing intensity. Nutrient dynamics in fertilised fish ponds are not well understood, and over-fertilisation can therefore occur. Fertilised ponds are usually static water systems with little exchange of water, so the risk of leakage of highly nutrient enriched water to the surrounding open waters is minimal (Pillay 1992).

The fertilisation of shrimp ponds may contribute to the nutrient load in surrounding receiving waters, although such effects have not been quantified (Phillips *et al.* 1993). No increased nutrient concentration was found in the discharge water from marine shrimp ponds enriched with cattle manure (Lee *et al.* 1986).

3.2.3 Residues of chemicals

In general, the use of chemicals to control diseases (bactericides, fungicides, parasiticides), aquatic vegetation (algicides, herbicides) and other nuisance organisms (insecticides, piscicides, molluscicides) in aquaculture, is required. Chemicals also include compounds to reduce handling trauma to organisms (anaesthetics) and to induce spawning or promote growth (hormones). There is also a range of compounds to disinfect water, improve water quality and increase productivity (lime, fertilisers).

National laws and regulations on the use of chemicals in aquaculture vary (Schnick 1991), especially in North America, Europe and Japan. Regulations are based on information regarding the efficacy of a chemical against its target organisms, toxicity to other organisms, human safety, residues in food-producing organisms and effects on the environment. Chemical treatments used in tropical aquaculture are often less regulated by legislation and the range of therapeutants commonly used appears to be more limited than in temperate countries (Beveridge & Phillips 1993).

A wide range of chemicals is used in aquaculture, but there are few data on the quantities used (Beveridge *et al.* 1991). Complete, reliable surveys of the quantities of chemicals and drugs used in Norwegian aquaculture are, however, available (Braaten 1992).

Chemotherapeutants

Antibiotics. The use of antibiotics is especially associated with intensive production, such as for salmonid and shrimp culture (Table 3.5). The following formulations were used in Norwegian salmonid production as feed-added antibacterial drugs during the period 1988–93: oxytetracycline chloride (OTC), nifurazolidone, oxolinic acid (OA), trimethoprim + sulfadiazine and flumequin (Braaten 1992; A Markestad, pers. comm.). The use of antibiotics since 1980 reflects the disease history of mariculture (Weston, this volume, Chapter 6).

Feed administered chemicals enter the environment through leaching from pellets and faeces and as a result of feed losses. Up to 20% of OTC can be leached from the pellets in 15 min (Fribourgh *et al.* 1969, cited in Beveridge *et al.* 1991). The digestibility of OTC and OA, fed to rainbow trout, was 7–9% and 14–38%, respectively (Cravedi *et al.* 1987). Experiments have shown that some of the ingested OA is excreted in an unchanged active form via the urine or faeces (O.B. Samuelsen, unpublished data).

In practice, at least 70% of the total feed-supplied antibiotics in mariculture is likely to be lost to the recipient. A standard treatment of the salmon stock in a 15 m

Table 3.5 Survey of commonly used chemicals in the mariculture of salmonids in Norway (Braaten 1992), in tropical inland aquaculture (Beveridge & Phillips 1993) and in tropical marine shrimp culture (Phillips *et al.* 1993).

Type of chemical	Mariculture of salmonids	Tropical inland aquaculture	Tropical marine shrimp culture
Therapeutants			
Antibacterials	Five types of antibiotics	Potassium permanganate (antibiotics rarely used)	Eight types of antibiotics
Parasitic treatment, Fungicides	Diclorvos, trichlorfon, praziquantel, formalin, fenbendazol, malachite green (hatcheries)	Malachite green, formalin, potassium permanganate, Dipterex	Formalin, methylene blue
Algicides (antifoulants)	Copper oxide, silicon wax	—	Copper sulphate
Piscicides, molluscicides	Not used	—	Tea seed cake, rotenone, nicotine, endrin, organotins
Disinfectants	Formalin, sodium hypochlorite, sodium hydroxide, chloramine	Formalin	Sodium hypochlorite, benzalkonium chloride
Water and soil treatment chemicals	pH regulators (lime, hatcheries)	Lime, potassium permanganate	Lime, EDTA, zeolite, calcium carbide, potassium permanganate
Anaesthetics	Benzocaine, metacaine, chlorbutanol, carbon dioxide	—	—

cage, with a stocking density of $15 \, kg/m^3$, requires 25 kg of OTC (Smith 1992). The majority of this OTC used will leave the cage associated with particles impacting the sediments beneath the cages. Several authors have detected antibiotic residues in the local wild fish faunas (e.g. Samuelsen *et al.* 1992). Sediment impacts from antibiotics are described by Austin (1993). Since recent studies indicate that OTC and OA are almost non-degradable in sediments (Braaten 1992), leakage seems to be the main reason for the observed short-lasting residues of these compounds in deposits beneath fish cages (e.g. Bjørklund *et al.* 1991).

Antibiotics are commonly used in penaeid hatcheries and growout ponds. Broad spectrum antibiotics, such as chloramphenicol, are frequently used in the former facilities. This chemical is now strongly restricted or banned in some countries, owing to the human health hazard and the development of antibiotic resistance (Brown 1989; Primavera 1991).

Parasiticides, fungicides. Bath treatment is commonly employed to treat skin and gill problems. The chemical compound is added directly into the water within the rearing units (tanks, ponds), or the fish are transferred to a treatment unit (mariculture).

Formalin and malachite green are in widespread use in hatcheries and land-based farms. They are also used as disinfectants, and are toxic to algae and zooplankton in concentrations below normal treatment levels (Phillips *et al*. 1993). Formalin, unlike malachite green, is easily diluted in water (Braaten 1992). These compounds are, in penaeid shrimp ponds, considered to be potentially harmful to both the internal pond ecosystem and the surrounding natural waters (Castille & Lawrence 1986). In European salmonid hatcheries and land-based farms, the average use of malachite green and formalin was 0.2 and 0.7 kg/t fish produced, respectively (Alabaster 1982).

Dipterex, the organophosphorus insecticide trichlorfon, has been widely used since 1961 to control ectoparasites and predatory insects in tropical ponds (Beveridge & Phillips 1993). This compound is highly toxic to fish (Flores-Nava & Vizcarra-Quiroz 1988), but studies indicate that it is rapidly hydrolysed under tropical conditions (Pillay 1992).

The most frequently used agent for treating salmon lice is the organophosphorus compound dichlorvos. It is added to treatment enclosures in concentrations of 0.5–2 g/m^3. Dichlorvos is toxic to fish and is extremely toxic to insects and crustaceans (an LD50 of 0.1–4 µg/l, Braaten 1992). Dichlorvos is gradually degraded to non-toxic compounds. A complete degradation occurs after 4.5 days in sea water at pH 7.4 (A. Syvertsen, unpublished data). Metriphonate (trichlorfon), which was common before the use of dichlorvos, is less toxic, but was considered more of an environmental risk because of the larger quantities required to give a similar disease control effect (Horsberg *et al*. 1987). The potential environmental loading of toxic agents from dichlorvos treatment can be reduced to 1–4% of the total quantity using metriphonate (Braaten 1992). The newly introduced chemical compounds with low environmental impacts, such as hydrogen peroxide and pyrethrum, and the use of wrasse as cleaner fish, are of interest as alternatives of dichlorvos to control sea lice (Thomassen 1993; Kvenseth 1993).

Algicides. Copper-based antifoulants are widely used in marine mariculture, e.g. in Norway the total use of copper oxide in 1990 amounted to *c*. 0.7 kg/t of salmon production (Braaten 1992). The sediments close to cages have been found to be enriched with copper (Braaten 1991), but no harmful effects were reported. Copper sulphate, commonly used to control algal growth in shrimp growout ponds in dose concentrations of 0.2 mg Cu/l (Primavera 1991), is considered to have a minimal impact on the natural environment (Phillips *et al*. 1993).

Other chemicals

Chlorinated hydrocarbons, such as endrin and aldrin, and organotins used as piscicides/molluscicides in shrimp ponds, are highly toxic, persistent compounds and pose a threat to the shrimp culture itself, as well as the surrounding environment (Phillips *et al.* 1993). Tea seed cake, nicotine and rotenone are biodegradable compounds which are expected to be less harmful.

Lime is routinely added to fish and shrimp ponds to increase the pH of the pond soil and water. In freshwater and brackishwater shrimp ponds, agricultural limestone is commonly applied at 500–1000 kg/ha to increase alkalinity (Boyd 1992). Burnt or hydrated lime is often applied at 1000–2000 kg/ha between crops, to sterilise pond bottoms. As much as 2–3 t/ha can be used in intensive fish ponds during the production cycle (Beveridge & Phillips 1993), usually without any adverse effects on the environment.

3.3 WASTE LOADINGS FROM DIFFERENT AQUACULTURE SYSTEMS

3.3.1 Intensive aquaculture systems

Most systems used for the production of fin fish in developed countries, i.e. rearing of salmonids, catfish, sea bream, sea bass and yellowtail, may be considered intensive. In developing countries intensive systems include freshwater and marine ponds (shrimps, carnivorous fish such as catfish and snakeheads), freshwater and marine cages and the pen culture of carnivorous fish (groupers, sea bass etc.) and some omnivorous fish, such as common carp (Pullin 1989). Though still amounting to a minor part of the total production in developing countries, intensive aquaculture in fresh water is starting to expand: for example in Chile (salmonids), Costa Rica (tilapia), China (grass carp and eel) and in Thailand (snakehead, walking catfish, giant freshwater prawns) (Martinez-Espinosa & Barg 1993).

A wide range of rearing systems is used for the intensive cultivation of fish and shellfish. These range between large, advanced, open-sea, floating systems, to simple, locally-made pens and enclosures in lakes and reservoirs. Land-based systems used include traditional ponds and more-recently-developed raceways, tanks and systems based on reuse of water.

Salmonids

The main farming system for the production of Atlantic and Pacific salmon is a first stage production in land-based hatcheries and smolt farms, using tanks and raceways, followed by a sea-based ongrowing production in floating cages. In Scotland, however, approximately half of the production of both Atlantic salmon smolts and rainbow trout is conducted in cages in freshwater lochs (NCC 1990). In north-west Europe, a small number of the ongrowing farms, 2–4% of the total, are land-based tank systems supplied with pumped sea water. Earth ponds, raceways and tanks are widely used for ongrowing rainbow trout, but cage farming in lakes (e.g. Scotland and Finland), brackish water and sea water (e.g. Norway and Canada) is common in some countries.

Land-based and closed salmonid culture systems are considered to have several environmental advantages and reduced waste loading, compared with floating cage systems (Eikebrokk *et al.* 1991):

- improved feed utilisation, because of reduced feed loss
- reduced consumption of antibiotics and chemicals (e.g. Dichlorvos);
- efficient removal of dead fish;
- reduced/eliminated risk of fish escapes;
- outlet wastewater treatment potential.

A wide range of specific waste loadings from freshwater salmonid farms in different European countries was demonstrated in a survey (NCC 1990). Reported waste loading rates are related to fish stock (g/kg fish/day), fish production (kg/t fish/year), or feed consumption (g/kg feed/day). In terms of kg/t fish produced the following ranges were quoted (mainly referring to rainbow trout culture): 289–2153 kg total suspended solids, 83–104 kg TN, 37–180 kg TAN, 11–110 kg TP and 2–8 kg dissolved P (DP). These loading rates refer mainly to studies conducted in the late 1970s and 1980s, during which time the high rates were strongly affected by feed loss. Improved feed quality and feeding procedures have decreased the mean FCR in both Norwegian salmon farming and Danish rainbow trout farming in 1990, to *c.* 1.35 (Eikebrokk *et al.* 1991; Jensen 1991).

The main pathways of organic matter and nutrients in a salmonid cage, assuming a high utilisation of energy enriched feed, are shown in Fig. 3.4. Approximately 23% of the BOD, 52% of the N and 63% of the P of the total supplied feed compounds load the environment. This amounts to 360 kg O_2 as BOD, 32 kg N and 7 kg P, per t of produced fish. The proportion of the total loading going to the sediments is highly dependent on the depth beneath the cages and the dominating current conditions.

In smolt farms high potential pollutant concentration peaks in the outlet water are especially related to the routine cleaning operations of the fish tanks (Table 3.6). Modern designed tanks for ongrowing have effective self-cleaning properties and/or sediment traps with a separate outlet, which produce fairly stable concentration levels. The highest specific loadings, at Scottish farms, are mainly related to the start-feeding period of fry with a high degree of overfeeding. Thus, the outlet loadings from hatcheries and smolt farms are both fluctuating throughout the day and seasonally (e.g. Hennessy *et al.* 1996).

The physical properties of the outlet compounds from flow-through systems influence the choice of suitable treatment strategies (Cripps 1993), e.g. the distribution of dissolved and particulate fractions of organic matter and nutrients (Cripps 1992). The potential for the mechanical removal of N is limited because of the high dissolved ammonia fraction. P is, however, more particle-based and so has potentially a higher removal rate (Cripps 1992; Bergheim *et al.* 1993; L.A. Kelly, pers. comm.).

Outside Norway, there are few reliable figures of the consumption of chemicals and

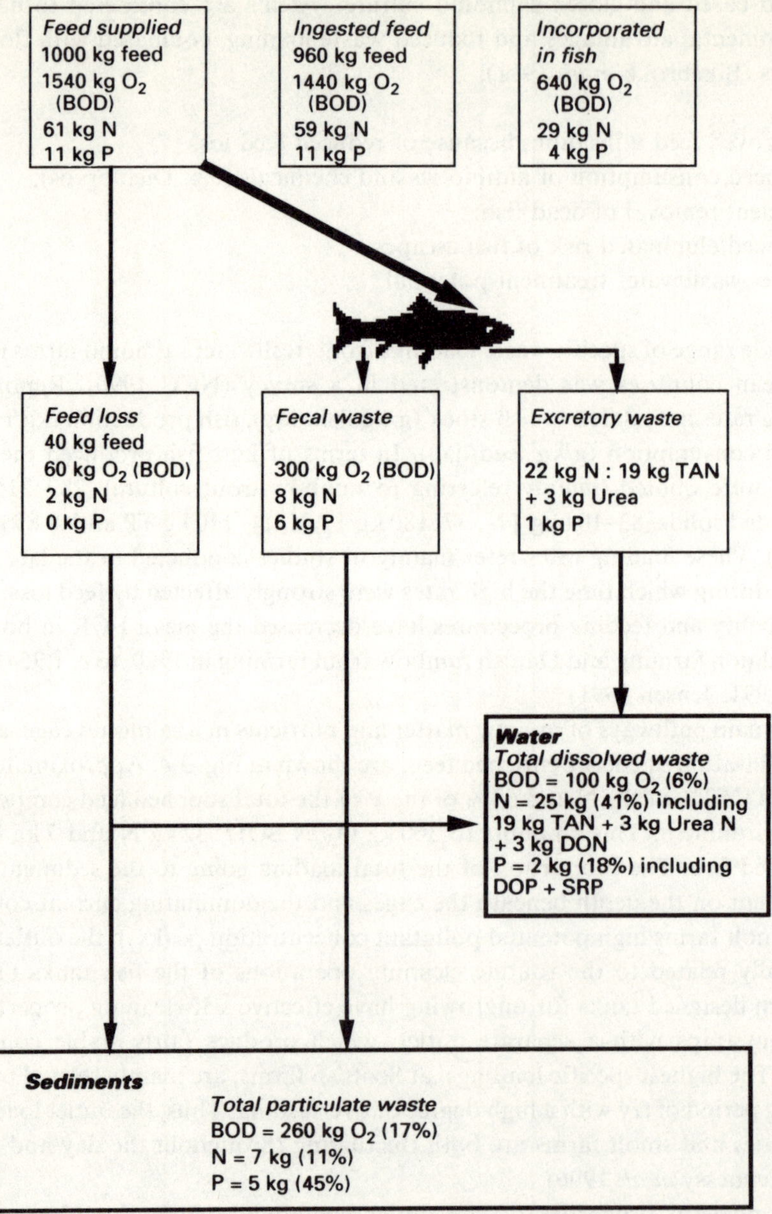

Fig. 3.4 Mass balance of organics and nutrients in a salmonid cage farm supplied with high energy feed. Mainly based on figures of Johnsen *et al.* (1993) and Håkanson *et al.* (1988).

Assumptions:

Feed quality: 38% protein, 30% fat, 23.4 MJ/kg wet weight. FCR: 0.96 kg/kg weight gain, 4% feed loss. Spent energy for basal metabolism, activity and digestion equivalent to 550 kg O_2 as BOD_5. Released to water from sinking pellets and fecal waste: 30% of N, 17% of P.

Table 3.6 Outlet concentrations and specific loadings of feed-derived compounds; smolt farms and a land-based ongrowing farm for Atlantic salmon. Scottish smolt farms (Hennessy *et al*. 1991); Norwegian smolt and ongrowing farm (Bergheim *et al*. 1991).

Parameter	Scottish smolt farms		Norwegian smolt farm		Norwegian ongrowing farm	
	Outlet concentration mg/l	Specific loadings g/kg fish/day	Outlet concentration mg/l	Specific loadings g/kg fish/day	Outlet concentration mg/l	Specific loadings g/kg fish/day
SS	0–201	0.98–36	1.0–11.3	0.62–11.7	0.60–1.9	0.58–2.5
BOD_5	0–181	1.1–549	—	—	—	—
TOC	—	—	0.30–4.3	0.14–2.2	0.40–0.60	0.33–0.79
Total N	0.004–2.8	0.17–7.2	0.30–0.93	0.17–0.81	0.31–0.38	0.29–0.49
TAN	0–1.6	0.07–5.5	0.08–0.17	0.05–0.09	0.12–0.17	0.15–0.18
TON	0–0.23	0–0.46	—	—	—	—
Total P	0–0.90	0.03–1.9	0.04–0.29	0.03–0.16	0.04–0.09	0.05–0.11

SS = suspended dry matter, BOD_5 = biochemical oxygen demand (5 days), TAN = total ammonia N, TON = total organic N, TOC = total organic carbon.

drugs (Weston, this volume, Chapter 6). Based on information from fish farmers, the use of antibiotics (active substance) per t of production in 1989 was 180 g in Denmark and 750 g in the Faeroe Islands, while the projected use in Scotland was approximately 400 g/t of production in 1986–1988 (Braaten 1991).

The history of the use of antibiotics reflects the disease infestation history of Norwegian aquaculture. A consumption peak occurred in 1987 owing to the outbreak of the so-called cold water vibriosis, when the salmonid biomass theoretically went through 1.3 standard antibacterial treatments annually (Grave *et al.* 1990). On average, the consumption of antibiotics in 1987 was *c.* 450 g active substance per t of fish produced. This was reduced to *c.* 100 g per t in 1989, owing to the development of an efficient vaccine against cold water vibriosis (Lillehaug 1990). The outbreaks of furunculosis caused a new peak of *c.* 200 g active substance per t in 1990–1992. This was again strongly reduced in 1993, mainly owing to the development of a successful vaccination programme (Norwegian Medicinal Depot, A. Markestad, pers comm.). Last year the use corresponded to only *c.* 45 g active ingredient per t of fish produced. Oxolinic acid and flumequin were most commonly used.

The consumption of both metriphonate and dichlorvos has gradually decreased during the 1990s because of recently introduced environmentally friendly methods for sea lice treatment (e.g. wrasse, hydrogen peroxide and pyrethrum), and general prophylactic measures to avoid lice outbreaks (e.g. reduced fish density and fallowing) (A. Markestad, pers. comm.).

Other fish species

The waste production from intensive farming systems producing non-salmonids has also gradually declined owing to improved feed quality. The development of Japanese yellowtail production is a typical example (Watanabe 1991). In the early 1980s the introduction of a moist pellet diet reduced the N and P loadings by about 50% compared with the previous loading levels using raw fish. Moreover, a new type of soft-dry pellet was developed in 1989 which resulted in the reduction in the excretion of N and P by a further 25% and 18%, respectively.

In intensive fish ponds, approximately 20–30% of the total feed supplied nutrients (commercially prepared feed) are used for fish growth:

- carp ponds, 11% of N, 32% of P, 69% of organic C; (Avnimelech & Lacher 1979)
- channel catfish ponds, 26.8% of N, 30.1% of P, 25.5% of organic matter (COD) (Boyd 1985)
- gilthead sea bream ponds, 26% of N, 21% of P (Krom & Neori 1989)

These mass budgets indicate that 70–90% of the total nutrient input is released into the water. The fraction actually loading the surrounding waters is, however, strongly dependent on the pond system (e.g. pond design and hydraulic loading). In the sea bream-stocked ponds (Krom & Neori 1989) the retention time of circularly moving sea water was 2 days and settled particles were daily drained out of the system. The measured outflow of nutrients balanced well with the input and fish-assimilated quantities.

In pond systems with low flow-through rates and long retention times, such as channel catfish ponds, the feed supported nutrient pathways will be strongly influenced by planktonic growth, bacterial activity and sedimentation (Table 3.7). During a growout period of 190 days, as much as 57% of the N input was released through denitrification and ammonia diffusion, while 55% of the P input was trapped in the bottom sediments. Assuming that proper sludge removal techniques were employed, less than 20% of the total nutrient waste, 18% of N and 15% of P, were loading the aquatic recipient. The aquatic loading corresponded, in terms of specific loading rates, to c. 9 g N and 1 g P per kg fish produced. This is considerably lower than the loading rates from single flow-through systems.

Table 3.7 Average gains and losses (kg) for nitrogen, phosphorus and chemical oxygen demand (COD) in channel catfish ponds. Earthen ponds (surface area $405\,m^2$, average depth 79 cm) stocked with 0.8–0.9 fish/ m^2 (Modified from Boyd 1985.)

Item	Nitrogen	Phosphorus	COD
Gain			
Feed	11.2 (92%)	1.68 (96%)	236 (36%)
Photosynthesis			413 (63%)
Other (fish stock,			11 (1%)
inflow, runoff, rain)	0.9 (8%)	0.07 (4%)	
Total	12.1	1.75	660
Losses			
Fish harvest	3.0 (25%)	0.52 (30%)	60 (9%)
Respiration:			
Water column			309 (47%)
Benthic			104 (16%)
Fish stock			121 (18%)
Denitrification/			
ammonia diffusion	6.9 (57%)		
Uptake by mud	0	0.97 (55%)	0
Overflow/		0.26 (15%)	66 (10%)
drainage/seepage	2.2 (18%)		
Total	12.1	1.75	660

Shrimps

In shrimp hatcheries and growout ponds, antibiotics are commonly used. There is no doubt that some antibiotics are used prophylactically to prevent outbreaks of disease, a practice which currently exposes the environment to risk (Phillips *et al* 1993). Indiscriminate use of antibiotics was considered to be the major factor in the collapse of the shrimp crop in Taiwan in 1988 that led to long-term resistance in the pathogens of the shrimp (Lin 1989).

Chemicals others than drugs, applied for pond preparation between crops, are also discharged from shrimp ponds. Toxic levels of formalin and teaseed powder can occur in adjacent waters during drainage of pond water soon after application (Primavera 1991).

Budgets of N and P were determined from comprehensive studies of intensive marine shrimp ponds in Southern Thailand (Briggs & Funge-Smith 1994, Table 3.8). The water exchange rates were 0.2–2.3%/day and the pond sediments were retained between harvests. On average 78–79% of the N and 92–95% of the P added to the ponds directly loaded the environment. A major portion of N (31%) and P (84%) was retained in the sediments, whilst the water-borne nutrients amounted to c. 35% and c. 10% of added N and P, respectively.

Table 3.8 Nutrient budgets of intensive marine shrimp ponds in Thailand (after Briggs & Funge-Smith 1994). FCR: 2.0 kg feed/kg shrimp produced. Units: percentage of total.

Input/output pond	Nitrogen (%)	Phosphorus (%)
Input		
Water	3.9–5.8	2.1–2.4
Fertilization	1.4–3.5	4.6–35.0
Feed	90.5–94.7	37.7–60.3
Shrimp + rain + runoff	0.2–0.3	0.1–0.2
Output		
Water	20.1–25.1	6.0–7.1
Drainage	12.6–13.4	2.9–3.6
Seepage	0.05–0.14	0.02–0.03
Sediment	27.2–34.0	82.1–86.2
Other*	8.6–17.6	0
Environment	77.7–79.8	92.3–95.1
Shrimp	20.2–22.3	4.9–7.7

* denitrification + ammonia diffusion

Based on data from other authors, Phillips *et al.* (1993) suggested that 67–78% of N and 76–86% of P fed to shrimps (*P. monodon*) in intensive pond culture are totally lost to the environment (FCR 1.2–2.0). In terms of the total waste per t of shrimp produced, the estimated loads were 57.3–118.1 kg of N and 13.0–24.4 kg of P. The corresponding loads from Thailand ponds (Briggs & Funge-Smith 1994) were 102 ± 25 kg N and 46 ± 14 kg P per t of shrimp produced (FCR 2.0).

The part of the total loading of organic matter and nutrients from intensive shrimp ponds discharged into the external environment is, as in fish ponds, closely connected to the sludge management and water exchange. If the deposited sludge is frequently resuspended and moved, the majority of the fed N can be discharged into the recipient waters (Fig. 3.5). Using the more traditional management regime, leaving the sludge undisturbed until harvest, a significant part of the N is lost to the atmosphere, because of the denitrification processes in the sludge. Of the three sludge management regimes (Fig. 3.5), frequent sludge removal during the growth season appears to offer the most advantages because of improved water quality for shrimp production and reduced nutrient discharge to the environment (Hopkins *et al.* 1994).

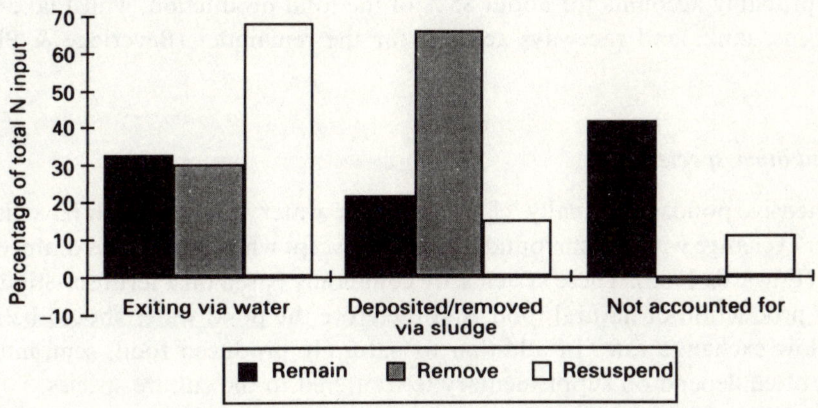

Fig. 3.5 Nitrogen waste pathways in intensive pond culture of shrimp at different management regimes of deposited sludge: (1) REMAIN, sludge left undisturbed in the pond; (2) REMOVE, sludge removed weekly; and (3) RESUSPEND, sludge resuspended or moved daily. (After Hopkins *et al.* 1994.)

'Not accounted for': mainly N released to the atmosphere (sludge denitrification).

Particulate material discharged from shrimp ponds can accumulate below the outlet point, especially in drainage and irrigation canals (Phillips *et al.* 1993). Harvesting from Thai shrimp ponds caused a substantial loss of particulate and organic matter, amounting to 44–48% of the total suspended solid discharged from the ponds during the production cycle (Briggs & Funge-Smith 1994).

Normally, effluents from shrimp ponds do not contain toxic substances (Boyd 1992). Measurable concentrations of reduced toxic substances in effluents, such as hydrogen sulphide, originating from disturbed bottom sludge, may occur during pond drainage at harvest. During periods of normal water exchange, the quality of the effluents will be similar to that of the pond water containing increased levels of nutrient and organic matter, and seldom settleable solids in measurable quantities.

Alternatively, high shrimp production can be achieved in round concrete ponds (Wyban & Sweeney 1989). A circular water movement can be established that effectively concentrates the sludge in the pond centre prior to flushing out through a central drain.

3.3.2 Semi-intensive systems

In developing countries, semi-intensive systems are currently of most importance, and this is likely to continue into the future (Pullin 1989). Such systems include fresh- and brackish-water pond farming of fish (e.g. carp and tilapia), shrimps and prawns, integrated agriculture/aquaculture pond systems, freshwater cage and net culture in eutrophic waters. Of the total freshwater fish production in the tropics, pond-based culture probably accounts for about 85% of the total production, whilst lake-based cages, pens, tanks and raceways account for the remainder (Beveridge & Phillips 1993).

Carp and other species

Semi-intensive ponds are usually 'closed' or static water systems which have little or no water exchange with the surrounding waters, except when the pond is drained after harvest (Edwards 1993). These systems are commonly based on a fertiliser-stimulated internal production of natural food, and therefore the pond water should be static, with a low exchange rate. In addition to naturally produced food, semi-intensive systems often depend on supplementary feed offered to the culture species.

As described above (3.2.2), nutrient conversion efficiencies (e.g. MCR) are lower in fertilised fish ponds than in intensive, feed based ponds. Work undertaken at the Asian Institute of Technology, cited by Edwards (1993), estimated that 170–491 g of N and 42–70 g of P was required to produce 1 kg of fish in some manure-fertilised ponds.

In terms of quantity of N and P released into surrounding waters, intensive systems are 7–31 and 3–11 times, respectively, more polluting than semi-intensive systems (Edwards 1993). These estimated relative ranges result from the use of different FCRs used to calculate the loading from intensive systems. In the example presented in Fig. 3.6, only 2% of the N and 6% of the P inputs were drained from semi-intensive ponds, whilst 83% of the N and 86% of the P were removed by the system, mainly trapped in the pond sediments. In practice, some additional nutrients retained in sediments would be flushed out during the pond-draining operation.

The therapeutants most commonly used in tropical inland aquaculture today are formalin, potassium permanganate, malachite green and Dipterex (trichlorfon) (Beveridge & Phillips 1993). Although it is generally assumed that the chemicals used are rapidly decomposed, there is little specific information regarding decomposition rates and breakdown products. According to Pullin (1989), semi-intensive freshwater ponds have no serious environmental effects other than their occupation of natural habitats.

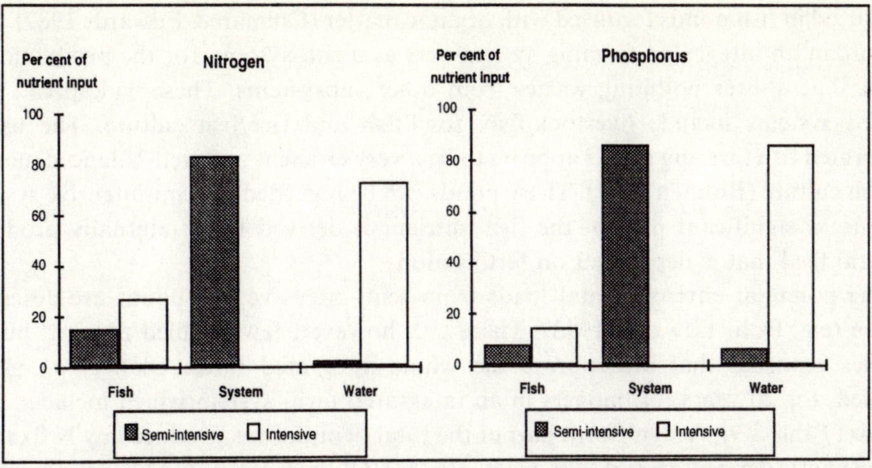

Fig. 3.6 A comparison of the nutrient budgets of a semi-intensive and an intensive aquaculture system with the same fish yield. Partitioning of added nitrogen and phosphorus, as percentage of those added in fertiliser (semi-intensive system) and feed (intensive system), assimilated in fish, disappeared in the system, and removed in water to pollute the external environment. After Edwards (1993).

Assumptions:

Semi-intensive system: pond stocked with Nile tilapia, fertilised with chicken manure + urea + triple superphosphate, only draining the system at the end of the culture cycle (static pond water). Intensive system: cage or raceway systems, supplied with complete feed, complete water exchange or flow-through.

Shrimps and prawns

The environmental effects from inorganic and organic fertilisation of shrimp ponds, such as those resulting from the nutrient load from the recipient, have not been quantified (Phillips *et al.* 1993). The nutrient budgets of the described shrimp ponds in Southern Thailand (Table 3.8) incorporate data from ponds to which both commercial feed and inorganic/organic fertilisers have been applied (Briggs & Funge-Smith 1994).

No increased nutrient loading from the discharge water from marine shrimp ponds was determined using an input of 1840 kg cattle manure/ha/week (Lee *et al.* 1986). The nutrient concentrations in the pond effluent were less than or equal to the nutrients in the incoming water. The authors concluded that a marine shrimp pond can be considered an efficient nutrient treatment plant converting the manure into valuable shrimp.

Integrated aquaculture systems

Integrated agriculture/aquaculture systems probably originated in China (Edwards 1993) and involve the reuse of farming waste, such as manure and crop residues, and human excreta, for fish culture. A large percentage of the global farmed fish crop is produced in fish ponds fertilised with organic matter (Colman & Edwards 1987). Fish culture in an integrated farming system acts as a sub-system, for the production of food, that utilises polluting wastes from other subsystems. These widespread integrated systems include livestock/fish, fowl/fish and rice/fish culture. The use of integrated fish farming ponds appears to be a very efficient and well-balanced method of fish culture (Braaten 1991). These ponds can be regarded as semi-intensive systems because a significant part of the fish nutrient is derived from internally produced natural feed that is dependent on fertilisation.

The potential environmental loads from semi-intensive fish ponds are described above (e.g. from Edwards 1993). There are, however, few detailed nutrient budget studies reported that incorporate the whole integrated farm. Oláh *et al.* (1992) studied, for 20 years, N budgets in an integrated farm system which included carp ponds (Table 3.9). A significant part of the total N input was produced by N fixation, atmospheric deposition and filling water (68%). Of the total input, 23% was removed by yield, 45% was retained by the sediments and 13% was removed by drainage water. The water outlet load from the ponds was less than the inflow load: 20 and 13% of the total N input, respectively. The authors concluded that when N-enriched water passes through integrated fish-cum-duck-alfalfa-rice systems, the farm operates as a natural filter, accumulating significant amounts of N in the pond bottom sediments.

3.3.3 Extensive systems

The environmental disadvantages of extensive systems, comprising coastal bivalve cultures, pen-cage cultures in eutrophic lakes, seaweed cultures and coastal fish

Table 3.9 Nitrogen budget during 20 years (1970–90) in a commercial fish-cum-duck-alfalfa-rice farm in Hungary (after Oláh *et al.* 1992). Units: t N/20 years, t N/ha/year and percentage of total input (%).

INPUT	Nitrogen	OUTPUT	Nitrogen (%)
Inorganic fertilization	212	Alfalfa yield	46 (4)
Duck feed	156	Fish yield	86 (7)
Fish feed	17	Rice yield	64 (5)
Manure	14	Duck yield	84 (7)
Atmospheric deposition	90	Draining water	170 (13)
Filling water	255	Denitrification	236 (19)
Alfalfa N fixation	210		
Fish pond N fixation	292		
Total input	1246	Total output	686 (55)
Total input/ha/year	0.207	Total output/ha/year	0.114
		Total input–output/ha/year	0.093

Assumptions: average N contents used, percentage: Manure 0.5, fish feed 1.6, duck feed 1.9, alfalfa 2.7, fish 2.5, rice 2.5, duck 2.7.

ponds, are mainly related to the occupation of natural ecosystems (Pullin 1989). Extensive fish ponds are characterised by a low fish density and high water consumption (Beveridge & Phillips 1993). These systems have no input of feed and fertiliser, and so the low waste production is considered to be of little environmental significance. The deposition of faeces and pseudofaeces, can, however, be considerable in mussel and oyster culture areas. An annual production of 1.76 g of fecal DM per g of mussel DM was found (Kautsky & Evans 1987), representing 130 mg C, 1.7 mg total N and 0.26 mg total P.

3.4 CONCLUSIONS

The quantity of the total waste from aquaculture leaving the system and loading the environment is closely connected to the culture system used. In intensive single-pass flow-through systems, such as tank-based hatcheries and floating cages, the waste produced continuously loads the surrounding waters. In systems with a low water exchange rate, such as semi-intensive carp ponds, a significant part of the waste produced will be removed within the system, mainly by accumulation in the pond sediments. Land-based and closed culture systems usually have several environmental advantages compared with floating cage systems.

There has been a well-documented reduction in the waste output of nutrients and organic matter from intensive culture systems during the last decade, due to better feed quality, improved feeding control and husbandry practice.

The use of antibiotics reflects the disease infestation history of aquaculture. The intensive cultures of salmonids and shrimps are associated with the high feeding rates. The development of efficient vaccines against vibriosis and furunculosis has greatly reduced the use of antibiotics in the North Sea mariculture of salmon since the late 1980s. Chemicals for parasite treatment and disinfection are widely used in aquaculture throughout the world, but there is little and inadequate information regarding the types and quantities used. There is certainly a need for more reliable global information on the use of drugs and chemicals in aquaculture. Currently, Norway is the only country to have introduced a reliable system to report the trade in these compounds.

ACKNOWLEDGEMENTS

Preparation of this paper was mainly sponsored by the Research Council of Norway within the Programme 'Closed production systems on land and at sea'. The authors are very grateful to Dr Simon Cripps (RF) for his constructive criticism and language correction. We would also like to thank Dr D. Seenappa, University of Bangalore, India, for his advice in preparing this manuscript.

REFERENCES

Akiyama, D.M. (1991) *The Use of Soy Products and Other Plant Protein Supplements in Aquaculture Feeds*. The American Soybean Association.

Alabaster, J.S. (1982) Survey of fish-farm effluents in some EIFAC countries. In: *Report of the EIFAC Workshop on Fish-farm Effluents* (ed. J.S. Alabaster), pp. 5–20. *EIFAC Technical Paper*, **41**. Silkeborg, Denmark.

Alanärä, A. (1992) The effect of time-restricted demand feeding on feeding activity, growth and feeding efficiency in rainbow trout (*Oncorhynchus mykiss*). *Aquaculture*, **108**, 357–68.

Alsted, N. (1991) Studies on the reduction of discharges from fish farms by modification of the diet. In: *Nutritional Strategies and Aquaculture Waste* (eds C.B. Cowey & C.Y. Cho), pp. 77–90. *Proceedings of the First International Symposium on Nutritional Strategies in Management of Aquaculture Waste*. University of Guelph, Guelph, Ontario, Canada, 1990.

Alsted, N.& Jokumsen, A. (1989) The influence of dietary protein on the growth of rainbow trout. In: *The Current State of Fish Nutrition in Aquaculture* (eds M. Takeda & T. Watanabe), pp. 209–20. *Proceedings of the Third International Symposium on Feeding and Nutrition in Fish*. Toba, Japan, 28 August–1 September.

Arnesen, P. (1993) Various carbohydrate feedstuffs in diets of Atlantic salmon (*Salmo salar* L.) in rainbow trout (*Oncorhynchus mykiss*, Walbaum). Dr Scient. thesis. Department of Animal Science, Agriculture University of Norway, Ås.

Åsgård, T. (1988) Nutritional value of animal protein sources for salmonids. In: *Proceedings Aquaculture International*, pp. 411–18. Vancouver, BC, Canada.

Åsgård, T., Langåker, R., Shearer, K.D., Austreng, E. & Kittelsen, A. (1991) Ration optimization to reduce potential pollutants – Preliminary results. *American Fisheries Society Symposium*, **19**, 410–16.

Ash, R. (1985) Protein digestion and absorption. In: *Nutrition and Feeding in Fish* (eds C.B. Cowey, A.M. Mackie & J.B. Bell), pp. 69–93. Academic Press, London.

Austin, B. (1993) Environmental issues in the control of bacterial diseases of farmed fish. In: *Environment and Aquaculture in Developing Countries* (eds R.S.V. Pullin, H. Rosenthal & J.L. Maclean), pp. 237–51. *ICLARM Conference Proceedings*, **31**.

Austreng, E. (1978) Digestibility determination in fish using chromic oxide marking and analysis of contents from different segments of the gastrointestinal tract. *Aquaculture*, **13**, 265–72.

Austreng, E. & Åsgård, T. (1991) Fôring av fisk i harmoni med miljøet. In: *Miljøhåndbok for fiskeoppdrett* (ed. J.I. Vikan), pp. 35–46. Aqua Books, Oslo, Norway.

Austreng, E., Skrede, A. & Eldegard, Å. (1980) Digestibility of fat and fatty acids in rainbow trout and mink. *Aquaculture*, **19**, 93–5.

Avnimelech, Y. & Lacher, M. (1979) A tentative nutrient balance for fish ponds. *Bamidgeh*, **31**, 3–8.

Bergheim, A. & Sveier, H. (1993) Redusert foforinnhold i tørrför til laksefisk, Del 2. Utprøving av fôr. *Report from Rogalandsforskning*, 114/93.

Bergheim, A., Tyvold, T. & Seymour, E.A. (1991) Effluent loadings and sludge removal from landbased salmon farming tanks. Paper for *AQUACULTURE EUROPE '91: Aquaculture and the Environment*. Dublin, Ireland.

Bergheim, A., Sanni, S., Indrevik, G. & Hølland, P. (1993) Sludge removal from salmonid tank effluent using rotating microsieves. *Aquacultural Engineering*, **12**, 97–109.

Beveridge, M.C.M. & Phillips, M.J. (1993) Environmental impact of tropical inland aquaculture. In: *Environment and Aquaculture in Developing Countries* (eds R.S.V. Pullin, H. Rosenthal & J.L. Maclean), pp. 213–36. *ICLARM Conference Proceedings*, **31**.

Beveridge, M.C.M., Phillips, M.J. & Clarke, R.M. (1991) A quantitative and qualitative assessment of wastes from aquatic animal production. In: *Aquaculture and Water Quality. Advances in World Aquaculture*. Vol. 3 (eds D.E. Brune & J.R. Tomasso), pp. 506–33. The World Aquaculture Society, Baton Rouge.

Bishara, N.F. (1979) Fertilizing fish ponds III. Growth of *Mugil capito* in Egypt by pond fertilization and feeding. *Aquaculture*, **16**, 47–55.

Bjørklund, H.V., Råbergh, C.M.I. & Bylund, G. (1991) Residues of oxolinic acid and oxytetracycline in fish and sediments from fish farms. *Aquaculture*, **97**, 85–96.

Blyth, P.J. & Burser, P. (1993) Detection of the feeding behaviour of seacaged Atlantic salmon, *Salmo salar*, using new 'adaptive' feeder technology. In: *Proceedings of the 6th International Symposium on Fish Nutrition and Feeding*. Hobart, Australia, 4–7 October 1993.

Boyd, C.E. (1976) Nitrogen fertilizer effects on production of *Tilapia* in ponds fertilized with phosphorus and potassium. *Aquaculture*, **7**, 385–90.

Boyd, C.E. (1985) Chemical budgets for channel catfish ponds. *Transactions of the American Fisheries Society*, **11**, 291–8.

Boyd, C.E. (1992) Shrimp pond bottom soil and sediment management. In: *Proceedings of the Special Session on Shrimp Farming* (ed. J. Wyban), pp. 166–81. World Aquaculture Society, Baton Rouge, LA, USA.

Braaten, B. (1991) Impact of pollution from aquaculture in six Nordic countries. Release of nutrients, effects, and waste water treatment. In: *Aquaculture and the Environment* (eds N. de Pauw & J. Joyce), pp. 79–101. *EAS Special Publication*, **16**. Gent, Belgium.

Braaten, B. (1992) *Forurensning fra Nordiske Akvakultur. Mengder, Effekt og Tiltak. NIVA Report*.

Braaten, B. & Hektoen, H. (1991) The environmental impact of aquaculture. In: *Report on a Regional Study and Workshop on Fish Disease and Fish Health Management. Fish Health management in Asia-Pacific*. ADB Agricultural Department Report Series No. 1, June 1991.

Briggs, M.R.P. & Funge-Smith, S.J. (1994) A nutrient budget of some intensive marine shrimp ponds in Thailand. *Aquaculture and Fisheries Management*, **25**, 789–811.

Brown, J. (1989) Antibiotics: their use and abuse in aquaculture. *World Aquaculture*, **20**, 35–43.

Buras, N. (1993) Microbial safety of produce from wastewater-fed aquaculture. In: *Environmental and Aquaculture in Developing Countries* (eds R.S.V. Pullin, H. Rosenthal & J.L. Maclean), pp. 285–95. *ICLARM Conference Proceedings*, **31**.

Castille, F.L. & Lawrence, A.L. (1986) The toxicity of erythromycin, minocycline, malachite green and formalin to nauplii of the shrimp *Penaeus stylirostris*. *Journal of the World Aquaculture Society*, **17**, 13–18.

Cho, C.Y. & Slinger, S.J. (1978) Apparent digestibility measurement in feedstuffs for rainbow trout. *Proceedings of the World Symposium on Finfish Nutrition and Fishfeed Technology*, **2**, 239–47.

Choubert, G. Jr., De la Noue, J. & Luquet, P. (1979) Continuous quantitative automatic collector for fish feces. *Progressive Fish Culturist*, **41**, 64–7.

Colman, J.A. & Edwards, P. (1987) Feeding pathways and environmental constraints in waste-fed aquaculture: balance and optimization. In: *Detritus and Microbial Ecology in Aquaculture* (eds D.J.W. Moriarty & R.S.V. Pullin), pp. 240–81. *ICLARM Conference Proceedings*, **14**.

Cowey, C.B. & Cho, C.Y. (eds) (1991) *Nutritional Strategies and Aquaculture Waste. Proceedings of the First International Symposium on Nutritional Strategies in Management of Aquaculture Waste*. University of Guelph, Guelph, Ontario, Canada, 1990.

Cowey, C.B. & Walton, M.J. (1989) Intermediary metabolism. In: *Fish Nutrition* (ed. J.E. Halver), pp. 259–329. Academic Press, London.

Cravedi, J-P, Choubert, G. & Delous, C. (1987) Digestibility of chloramphenicol, oxolinic acid and oxytetracycline in rainbow trout and influence of these antibiotics on lipid digestibility. *Aquaculture*, **60**, 133–42.

Cripps, S.J. (1992) Characterization of an aquaculture effluent based on water quality and particle size distribution data. *Abstract, Aquaculture '92 Conference*, 21–25 May 1992, Orlando, Florida, USA.

Cripps, S.J. (1993). The application of suspended particle characterization techniques to aquaculture systems. In: *Techniques for Modern Aquaculture* (ed. J-K Wang), pp. 26–34. *Proceedings of an Aquacultural Engineering Conference*, 21–23 June 1993, Spokane, Washington, USA.

Davenport, J., Kjorsvik, E. & Haug, T. (1990) Appetite, gut transit, oxygen uptake and nitrogen excretion in captive Atlantic halibut, *Hippoglossus hippoglossus* L., and lemon sole, *Microstomus kitt* (Walbaum). *Aquaculture*, **90** 267–77.

Delmendo, M.N. (1980) A review of integrated livestock–fowl–fish farming systems. In: *Integrated Agriculture Aquaculture Farming Systems* (eds R.S.V. Pullin & Z.H. Shehadeh), pp. 59–71. ICLARM Conference Proceedings, **4**.

Edwards, P. (1993) Environmental issues in integrated agriculture–aquaculture and wastewater-fed fish culture systems. In: *Environment and Aquaculture in Developing Countries* (eds R.S.V. Pullin, H. Rosenthal & J.L. Maclean), pp. 139–70. *ICLARM Conference Proceedings*, **31**.

Eikebrokk, B., Fløgstad, H., Hergheim, A. & Åsgård, T. (1991) Prospects and perspectives for the development of green production technologies in Northern Seas agriculture. In: *Seminar on Agriculture and Aquaculture* (eds J.M. Debois & B. Braaten), pp. 85–100. 1st ENS *Conference*, 26–30 August 1991, Stavanger, Norway.

Fivelstad, S., Thomassen, J.M., Smith, M.J., Kjartansson, H. & Sandø. A.-B. (1990) Metabolic production rates from Atlantic salmon (*Salmo salar* L.) and Arctic char (*Salvelinus alpinus* L.) reared in single pass land-based brackish water and sea-water systems. *Aquacultural Engineering*, **9**, 1–21.

Flores-Nava, A. & Vizcarra-Quiroz, J.J. (1988) Acute toxicity of Trichlorfon (Dipterex) to fry of *Cichlasoma urophthalmus* Gunther. *Aquaculture and Fisheries Management*, **19**, 341–410.

Forsberg, O.I. (in press). The impact of varying feeding regimes on oxygen consumption and excretion of carbon dioxide and nitrogen in post-smolt Atlantic Salmon (*Salmo salar*). *Aquaculture Research*.

Fribourgh, J.H., Meyer, F.P. & Robinson, J.A. (1969) Oxytetracycline leaching from medicated fish feeds. *US Bureau of Sport Fisheries and Wildlife Technical Paper*, **40**.

Grave, K., Engelstad, M., Søli, N.E. & Håstein, T. (1990) Utilization of antibacterial drugs in salmonid farming in Norway during 1980–1988. *Aquaculture*, **86**, 347–58.

Guérin-Aucey, O. (1976) Etude expérimental de l'excrétion azotée du bar (*Dicentrarchus labrax*) en cours de croissance. I. Effets de la température et du poids du corps sur l'excrétion. *Aquaculture*, **9**, 71–80.

Handy, R.D. & Poxton, M.G. (1993) Nitrogen pollution in mariculture: toxicity and excretion of nitrogenous compounds by marine fish. *Reviews in Fish Biology and Fisheries*, **3**, 205–41.

Helland, S.J. & Grisdale-Helland, B. (1993) A simple method for daily feed intake measurements for groups of fish. *Proceedings of the 6th International Symposium on Fish Nutrition and Feeding*. Hobart, Australia, 4–7 October 1993.

Hennessy, M., Wilson, L. & Struthers, W. (1991) *Management Strategies for Salmon Farm Effluents*. A report prepared for the Scottish Salmon Growers' Association, Perth.

Hennessy, M., Wilson, L., Struthers, W. & Kelly, L.A. (1996) Waste loadings from two Atlantic salmon juvenile farms in Scotland. *Water, Air and Soil Pollution*, **86**, 235–49.

Hillestad, M. & Johnsen, F. (1994) High energy/low protein diets for Atlantic salmon: effects on growth, nutrient retention and slaughter quality. *Aquaculture*, **124**, 109–16.

Hopkins, J.S., Sandifer, P.A. & Browdy, C.L. (1994) Sludge management in intensive pond culture of shrimp: Regime on water quality, sludge characteristics, nitrogen extinction, and shrimp production. *Aquacultural Engineering*, **13**, 11–30.

Horsberg, T.E., Berge, G.N., Djupvik, H.O., Hektoen, H., Hognestad, I.M. & Ringstad, R. (1987) Diklorvos som avlusningsmiddel for fisk. Klinisk utprøving og toksistetstesting. *Norsk Veterinårtidsskrift*, **9**, 611–15.

Håkanson, L., Ervik, A., Mäkinen, T. & Møller, B. (1988) *Basic Concepts Concerning Assessments of Environmental Effects of Marine Fish Farms*. Nordic Council of Ministers, Nord 1988, **90**.

Jensen, J.B. (1991) Environmental regulations of fresh water fish farms. In: *Nutritional Strategies and Aquaculture Waste* (eds C.B. Cowey & C.Y. Cho), pp. 251–62. *Proceedings of the First International Symposium on Nutritional Strategies in Management of Aquaculture Waste*. University of Guelph, Guelph, Ontario, Canada, 1990.

Jobling, M. (1993) Bioenergetics: feed intake and energy partitioning. In: *Fish Ecophysiology* (eds J.C. Rankin & F.B. Jensen), pp. 1–44. Chapman & Hall, London.

Johnsen, F. & Wandsvik, A. (1991) The impact of high energy diets on pollution control in the fish farming industry. In: *Nutritional Strategies and Aquaculture Waste* (eds. C.B. Cowey & C.Y. Cho), pp. 51–63. *Proceedings of the First International Symposium on Nutritional Strategies in Management of Aquaculture Waste*. University of Guelph, Guelph, Ontario, Canada, 1990.

Johnsen, F., Hillestad, M. & Austreng, E. (1993) High energy diets for Atlantic salmon. Effect on pollution. In: *Fish Nutrition in Practice* (eds S.J.Kaushik & P. Luquet), pp. 391–401. INRA, Paris.

Kaushik, S.J. (1980) Influence of nutritional status on the daily patterns of nitrogen excretion in the carp (*Cyprinus carpio* L.) and the rainbow trout. *Reproduction, Nutrition and Development*, **20**, 1751–65.

Kautsky, N. & Evans, S. (1987) Role of biodeposition by *Mytilus edulis* in the circulation of matter and nutrients in a Baltic coastal ecosystem. *Marine Ecology Progress Series*, **38**, 201–12.

Kiaerskou, J. (1991) Production and economics of 'Low pollution diets' for the aquaculture industry. In: *Nutritional Strategies and Aquaculture Waste* (eds C.B. Cowey & C.Y. Cho), pp. 65–76. *Proceedings of the First International Symposium on Nutritional Strategies in Management of Aquaculture Waste.* University of Guelph, Guelph, Ontario, Canada, 1990.

Knights, B. (1989) Effects of ammonia accumulation on metabolic rates and growth of European eel, *Anguilla anguilla* L., in relation to warmwater aquaculture. *Aquaculture and Fisheries Management,* **20,** 111–17.

Krom, M.D. & Neori, A. (1989) A total nutrient budget for an experimental intensive fishpond with circularly moving seawater. *Aquaculture,* **83,** 345–58.

Kvenseth, P.G. (1993) Use of wrasse to control salmon lice. In: *Fish Farming Technology* (eds H. Reinertsen, L.A. Dahle, L. Jørgensen & K. Tvinnereim), pp. 227–32. *Proceedings of the First International Conference on Fish Farming Technology.* Trondheim, Norway.

Lall, S.P. (1991) Digestibility, metabolism and excretion of dietary phosphorus in fish. In: *Nutritional Strategies and Aquaculture Waste* (eds C.B. Cowey & C.Y. Cho), pp. 21–36. *Proceedings of the First International Symposium on Nutritional Strategies in Management of Aquaculture Waste.* University of Guelph, Guelph, Ontario, Canada, 1990.

Lee, C.S., Sweeney, J.N. & Richards, W.K. (1986) Marine shrimp aquaculture: a novel waste treatment system. *Aquacultural Engineering,* **5,** 147–60.

Lillehaug, A. (1990) A field of trial vaccination against cold-water vibriosis in Atlantic salmon (*Salmo salar* L.). *Aquaculture,* **84,** 1–12.

Lin, C.K. (1989) Prawn culture in Taiwan: What went wrong? *World Aquaculture,* **20,** 19–20.

Lovell, R.T. (1989) Diet and fish husbandry. In: *Fish Nutrition* (ed. J.E. Halver), 2nd edn., pp. 550–605. Academic Press.

Ludwig, G.M. & Tackett, D.L. (1991) Effects of using rice bran and cottonseed meal as organic fertilizers on water quality, plankton, and growth and yield of striped bass, *Morone saxatilis*, fingerlings in ponds. *Journal of Applied Aquaculture,* **1,** 79–94.

Martinez-Espinoza, M. & Barg, U. (1993) Aquaculture and management of freshwater environments, with emphasis on Latin America. In: *Environment and Aquaculture in Developing Countries* (eds R.S.V. Pullin, H. Rosenthal & J.L. Maclean), pp. 42–59. *ICLARM Conference Proceedings,* **31.**

Meyer-Burgdorff, K.-H. (1992) Nutritional strategies to reduce phosphorus excretion of farmed fish. In: *The Workshop 'Fish Farm Effluents and their Control in EC Countries': Abstracts* (eds H. Rosenthal & V. Hilge), p. 60. Hamburg, Germany, November 1992.

NCC (1990) *Fish Farming and the Scottish Freshwater Environment.* Nature Conservancy Council, Edinburgh.

Ogino, C., Kakino, J. & Chen, M.-S. (1973) Protein nutrition in fish. II. Determination of metabolic fecal nitrogen and endogenous nitrogen excretion of carp. *Bulletin of the Japanese Society of Scientific Fisheries,* **35,** 519–23.

Oláh, J., Szabó, P., Esteky, A.A. & Nezami, S.A. (1992) Nitrogen processing and retention on a Hungarian carp farm. In: *The Workshop 'Fish Farm Effluents and their Control in EC Countries': Abstracts* (eds H. Rosenthal & V. Hilge), p. 43. Hamburg, Germany, November 1992.

Phillips, M.J., Kwei Lin, C. & Beveridge, M.C.M. (1993) Shrimp culture and the environment: lessons from the world's most rapidly expanding warmwater aquaculture sector. In: *Environment and Aquaculture in Developing Countries* (eds R.S.V. Pullin, H. Rosenthal & J.L. Maclean), pp. 171–97. *ICLARM Conference Proceedings,* **31.**

Piedrahita, R. & Tchobanoglous, G. (1987) The use of human wastes and sewage in aquaculture. In: *Detritus and Microbial Ecology in Aquaculture* (eds D.J.W. Moriarty & R.S.V. Pullin), pp. 336–52. *ICLARM Conference Proceedings,* **14.**

Pillay, T.V.R. (1992) *Aquaculture and the Environment.* Fishing News Books, Oxford.

Porter, C.B., Krom, M.D., Robbins, M.G., Brickell, L. & Davidson, A. (1987) Ammonia excretion and total N budget for gilthead seabream (*Sparus aurata*) and its effect on water quality conditions. *Aquaculture,* **66,** 287–97.

Primavera, J.H. (1991) Intensive prawn farming and the environment. *Paper presented at The Second Philippine Prawn Congress,* Bacolod City, Negros Occidental, No. 1991, 18 p.

Pullin, R.S.V. (1989) Third-world aquaculture and the environment. *NAGA, The ICLARM Quarterly,* **12,** 10–13.

Pullin, R.S.V., Rosenthal, H. & Maclean, J.L. (eds) (1993) *Environment and Aquaculture in Developing Countries. ICLARM Conference Proceedings,* **31.**

Ramnarine, I.W., Pirie, J.M., Johnstone, A.D.F. & Smith, G.W. (1987) The influence of ration size and feeding frequency on ammonia excretion by juvenile Atlantic cod, *Gadus morhua* L. *Journal of Fish Biology*, **31**, 545–59.

Samuelsen, O.B., Lunestad, B.T., Husevåg, B., Hølleland, T. & Ervik, A. (1992) Residues of oxolinic acid in wild fauna following medication in fish farms. *Diseases of Aquatic Organisms*, **12**, 111–19.

Sayer, M.D.J. & Davenport, J. (1987) The relative importance of the gills to ammonia and urea excretion in five seawater and one freshwater teleost species. *Journal of Fish Biology*, **31**, 561–70.

Schnick, R.A. (1991) Chemicals for worldwide aquaculture. In: *Fish Health Management in Asia-Pacific. Report on a Regional Study and Workshop on Fish Disease and Fish Health Management*, pp. 441–67. (ADB/NACA). ADB Agriculture Department Report Series, **1**. Bangkok, Thailand.

Smith, P. (1992) Antibiotics and the alternatives. *Aquaculture and the Environment* (eds N. dePauw & J. Joyce), pp. 223–34. *EAS Special Publication*, **16**. Gent, Belgium.

Steffens, W. (1989) *Principles of Fish Nutrition*. Ellis Horwood, Chichester.

Thomassen, J.M. (1993) A new method for control of salmon lice. In: *Fish Farming Technology* (eds H. Reinertsen, L.A. Dahle, L. Jørgensen & K. Tvinnereim), pp. 233–36. *Proceedings of the First International Conference on Fish Farming Technology*. Trondheim, Norway.

Watanabe, T. (1991) Past and present approaches to aquaculture waste management in Japan. In: *Nutritional Strategies and Aquaculture Waste* (eds C.B. Cowey & C.Y. Cho), pp. 137–54. *Proceedings of the First International Symposium on Nutritional Strategies in Management of Aquaculture Waste*. University of Guelph, Guelph, Ontario, Canada, 1990.

Wickins, J.F. (1985) Ammonia production and oxidation during the culture of marine prawns and lobsters in laboratory recirculation systems. *Aquacultural Engineering*, **4**, 155–74.

Wohlfarth, G.W. & Hulata, G. (1987) Use of manures in aquaculture. In: *Detritus and Microbial Ecology in Aquaculture* (eds D.J.W. Moriarty & R.S.V. Pullin), pp. 353–65. *ICLARM Conference Proceedings*, **14**.

Wohlfarth, G.W. & Schroeder, G.L. (1979) Use of manure in fish farming – a review. *Agricultural Wastes*, **1**, 279–99.

Wohlfarth, G.W. & Schroeder, G.L. (1991) Potential benefits of manure in aquaculture: a note qualifying the conclusions from our paper on the dominance of algal-based food webs in fish ponds. *Aquaculture*, **94**, 307–8.

Wyban, J.A. & Sweeney, J.N. (1989) Intensive shrimp growout trials in a round pond. *Aquaculture*, **76**, 215–25.

Yang, H., Fang, Y. & Chen, Z. (1992) Integrated fish farming systems in China and the allocation of resources. *World Aquaculture*, **23**(1), 61–8.

Chapter 4
Environmental Impacts of Nutrients from Aquaculture: Towards the Evolution of Sustainable Aquaculture Systems

Barry A. Costa-Pierce   *Center for Regenerative Studies, California Polytechnic State University, California, USA*

4.1   INTRODUCTION

Scientific interest in the impacts of nutrient pollution from aquaculture has increased markedly since the 1980s (Rosenthal *et al.* 1987; ICES 1989; Institute of Aquaculture [IOA] *et al.* 1990; Cowey & Cho 1991; DePauw & Joyce 1991; Iwama 1991; Mäkinen 1991; Pullin *et al.* 1993), and especially so in the past 10 years as aquaculture has become one of the world's fastest growing agriculture industries (Davlin 1991). Concerns about aquaculture pollution, combined with real and perceived water quality degradation, health concerns, and other violations of the public trust, have fuelled vigorous public and policy debates. Controversies have resulted in adoption of regulations intended to preserve the integrity of natural ecosystems and to ameliorate the public's concerns. The increasing regulatory burden has been cited as one of the main factors slowing the growth of aquaculture (Axler *et al.* 1994).

The political strength of international and grass roots environmental organisations, combined with the ascent to power of a post-World War II generation acutely aware of the potential environmental impacts of food-producing industries have dramatically changed environmental policies world-wide. Regulations have been promulgated to address new concerns about the preservation, sustainability and stewardship of the environment. As a result, the water pollution control agenda has shifted from a 'user pays' to a 'preservationist' stance that opposes addition of any new sources of aquatic pollution, combined with a switch in management philosophy to adoption of sustainability, and integrated and ecosystems criteria. Holistic water systems management, and watershed-level planning processes have become modern dogma in the attempt to manage natural environments better. However, the methods needed for such ecosystem level approaches remain poorly developed.

Such concepts are the wave of the present and the foundations for the future in water pollution control and environmental management. Clearly, the growth of aquaculture could be seriously impeded if it confronted or disregarded these trends. The damaged and degraded state of most aquatic ecosystems world-wide combined with public concerns about adding any 'new' sources of pollution to already overburdened natural ecosystems will require aquaculture to develop ecosystems approaches and sustainable operating procedures, and to articulate a sustainable, ecological aquaculture systems pedagogy.

In the twenty-first century aquaculture developers will need to spend as much time on the technological advances coming to the field as they do on designing ecological approaches to aquaculture development that clearly exhibit stewardship of the environment. For aquaculture development to proceed to the point where it will be recognised world-wide as a major contributor to new agriculture production, clear, unambiguous linkages between aquaculture and the environment must be created and fostered, and the complementary roles of aquaculture in contributing to environmental sustainability, rehabilitation and enhancement must be developed and clearly articulated to a highly-concerned, increasingly-educated and involved public.

4.2 AQUACULTURE SYSTEMS AND OPERATIONAL INTENSITIES

Aquaculture systems can be classified by their location within or separate from the environment, and by the levels of operational intensity (management, feeds, water flows, etc.) and integration (Table 4.1). The most common type of modern aquaculture system according to any type of modern classification scheme is earthen ponds growing omnivorous fish species produced in hatcheries, being fed supplemental feeds on an exact feeding schedule. Most of these pond systems are located in Asia, and are open systems having no investments in waste treatment facilities of any kind.

Not unexpectedly, aquaculture systems that have a high intensity of production and discharge wastes with no treatment whatsoever to oligotrophic ecosystems, have the greatest potential for nutrient impacts on the environment. It is assumed here that the amount of dependence on the natural environment for waste assimilation is directly related to the amount of on-site treatment of wastes that is performed. While super-intensive, flow-through aquaculture systems have potentially the highest nutrient impact on natural ecosystems, impacts from such systems can be insignificant if complete, on-site waste treatment occurs. Therefore, super-intensive aquaculture cannot always be assumed to have major nutrient impacts that impair natural aquatic ecosystem structure and functions.

Any such classification of aquaculture systems based upon nutrient impacts alone is incomplete and can be used as a guide only, since most modern aquaculture systems having no environmental impact are generally not commercially viable. For example, a closed loop tank facility growing herbivorous fish on natural foods is an economic impossibility. However, summarising the recent data reviewed in this paper, it is possible to generalise that the nutrient impacts of floating cage and raft aquaculture systems are of greater environmental concern than intensive recirculating systems, and that semi-intensive pond systems are of little environmental concern.

Table 4.1 Classifications of aquaculture systems.

Types	Kinds and levels
Stocking, management and economic intensity levels	Intensive
	Semi-intensive
	Extensive
Water salinities	Freshwater
	Brackishwater
	Seawater
Water flow characteristics	Running water (lotic)
	Standing water with flushing
	Standing water (lentic)
Amount of on-site waste treatment and recirculation	Open, no recirculation
	Semi-closed, partial recirculation
	Closed, full recirculation
Environmental location	Indoor
	Outdoor-natural
	Outdoor-artificial
Feed qualities	Complete
	Supplemental
	Natural
Feeding strategies	Continuous
	Scheduled
	Natural
Species stocking strategies	Monoculture
	Janitorial polyculture
	Polyculture
Species temperature tolerances	Eurythermal
	Stenothermal
	Cold water
	Warm water
Species salinity tolerances	Euryhaline
	Stenohaline
	Marine (mariculture)
	Brackishwater
Species natural food habits	Carnivorous
	Omnivorous
	Herbivorous
	Opportunistic
Fry sources	Hatcheries
	Wild capture of broodstock
	Natural
Level of systems integration	Stand alone
	Integrated
Unit types	Raceways
	Tanks
	Cages (floating, fixed)
	Net pens (fixed)
	Rafts (ropes, nets)
	Ponds
Marketing channels	Human food (local, export)
	Sport, recreational tourism

4.3   CHARACTERISATION OF AQUACULTURE NUTRIENTS

The majority of modern aquaculture operations do not completely treat their wastes (e.g. there is nowhere a completely closed loop). Therefore, natural ecosystems, terrestrial and aquatic, are used as an integral part of the waste treatment process for aquaculture systems. Solbé (1982) reported that only 17% of farms in the UK treat their wastes; UMA (1988) reported that only 26% of farms in Ontario, Canada, treated their effluents. In both of these, and in other systems (see Westers 1991), those farms that do practise any kind of waste treatment use settling ponds only.

If aquaculture systems are located completely within natural environments (cages, rafts) and are not required to treat their wastes, the public is subsidising the environmental costs of these operations at the level of (1) the additional capital costs for complete waste treatment, (2) the operating costs for treatment facilities, and (3) the interest on loans received to purchase and operate waste treatment systems (Westers 1991). This assumes that aquaculture wastes can be treated completely and that any aquatic environmental changes due to aquaculture wastes are undesirable.

There have been a number of studies and reviews that have measured and summarised the solid and nutrient wastes emanating from intensive aquaculture systems in the temperate zone. A summary of nutrient discharges is given in Table 4.2. Examination of the data reveals the potential for nutrient impact on aquatic ecosystems if these effluents are discharged to enclosed basins, to natural systems with low flushing rates, or to vulnerable ecosystems having species of special concern.

Table 4.2   Nutrient discharges from intensive aquaculture reported by various authors (kg/t fish ± SD).

Culture system	Species	Country	TN	TP	TN/TP
Ponds, tanks	Trout	Ireland	83	11	7.5
Cages	Trout	Poland	97	23	4.2
Cages	Trout	Sweden	87	13.5	6.4
Cages	Trout	Scotland	104	27	3.8
Cages	Yellowtail	Japan	68	23	3.0
Cages	Yellowtail	Japan	109	28	3.9
Cages	Bream	Japan	211	62	3.4
Mean			108.4	16.6	4.7
Standard deviation			47.3	16.5	1.8

Solid wastes from aquaculture consist of waste feed (fines and uneaten feed) and fish feces. Available data indicate that about 0.5 t of settleable solids is produced for every tonne of fish produced (Table 4.3). There are, however, major exceptions, Solbé (1987) reported 1.35 t of solids produced for every tonne of fish in a survey of freshwater fish farms in the UK in 1980. Any summary of available studies of nutrient releases masks the variability in solid waste discharges that have been shown to vary according to environmental and production conditions such as seasonality, water

Table 4.3 Summary of the composition of aquaculture effluents reported by various authors.

Effluent component	No. of studies	kg/t fish	Authors
Temperate zone cage culture			
Settleable solids	9	510.7 ± 207.3	IOA *et al.* (1990)
TN	7	108.4 ± 47.3	IOA *et al.* (1990); Watanabe (1991)
TP	17	19.2 ± 10.8	McDonald *et al.* (1994)
TN/TP (here)	7	5.6	
TN/TP (from Table 4.2)	6	4.7 ± 1.8	
Indonesian carp culture in cages			
Settleable solids		2.7 ± 1.8	Costa-Pierce & Roem (1990)
TP		90.2 ± 90.4	
TN		0.1 ± 0.2	
TN/TP		0.001	

quality in the aquaculture system and amount of fish stress, the quality, quantity and feeding schedule of formulated diets, efficiency of production and harvesting schedules (Boyd 1978; Bergheim *et al.* 1982; Beveridge 1987; IOA *et al.* 1990; Schwartz & Boyd 1994). However, while there is a large variability in the studies, mean nutrient concentrations of total nitrogen (TN) and total phosphorus (TP) are similar among many types of intensive systems. Also, it is important to note that the inorganic and organic fractions of TN and TP in these aquaculture wastes differ. The major part (*c.* 60–90%) of the TN in aquaculture wastes has been found in the dissolved fraction, whereas the major portion of the TP has been found in the organic, particulate fraction (*c.* 60–90%) (Enell & Lof 1983; Phillips 1985). Estimates range from 45 to 55.5 kg of inorganic nitrogen, primarily ammonium and urea, discharged for every tonne of fish produced (Gowen & Bradbury 1987).

4.3.1   Patterns in nutrient release from aquaculture

Variability in waste production has been studied for cage systems and semi-intensive pond aquaculture. The major proportion of solids output occurs during the warm season in temperate zone cages, or in 'pulses' of solids output during harvesting of semi-intensive ponds. IOA *et al.* (1990) calculated that solid loads from the aquaculture of Atlantic salmon smolts exhibited strong seasonal variation, with large pulses in solids output when warmer temperatures occurred and feeding rates increased in the spring and summer.

In earthen ponds and tanks, Bergheim *et al.* (1982) showed that tank cleaning can accelerate solids production by 100–200 times the routine output. Boyd (1978) demonstrated that effluents discharged during pond harvesting operations for channel catfish ponds in the USA were highly enriched in solids and nutrients in comparison with routine operations. Ellis *et al.* (1978) reported that settleable solids from fishponds exceeded the US EPA standard of 3.3 mg/l only 15% of the time, and that these values were during final harvesting. More recently, Schwartz & Boyd (1994) showed that effluents from catfish ponds contained solids and nutrients comparable to or lower than those of receiving waters during normal operations, but discharged 50% of the total suspended solids in the last 5% of the pond volume drained during fish harvesting.

The pulsed nature of pond outputs could work to the advantage of designers of aquaculture waste treatment systems since some measure of predictability is present. All that may be required is proper placement and sizing of settling ponds to capture the large amounts of wastes disproportionately discharged at harvests (Westers 1991; Schwartz & Boyd, 1994). In addition, it appears that routine operations of semi-intensive ponds produce environmentally insignificant amounts of pollutants both in temperate zone ponds (Boyd 1978; Schwartz & Boyd, 1994) and in the tropics (Edwards 1993).

The quality of aquaculture wastes and its temporal variability are important determinants of environmental impacts. One way of measuring nutrient quality, potential environmental impacts and trophic responses is evaluation of the TN/TP ratios (Downing & McCauley 1992; Seip 1994). Table 4.4 gives a list of TN/TP ratios for various nutrient sources, ranging from less than 0.1 to 247. A balanced, or Redfield, ratio for the TN/TP of phytoplankton is 7.2 (Redfield 1958). Calculation of the TN/TP ratios of wastes from intensive aquaculture for the seven data sets where both data exist gives a TN/TP ratio of 5.6 (Table 4.3), similar to the mean ratio of 4.7 ± 1.8 calculated from Table 4.2. Such a low ratio indicates that aquaculture wastes are highly enriched in TP. Sedimenting materials from intensive aquaculture are therefore comparable in composition to effluents from urban storm-water drainage, with a TN/TP of 5.8 (Loehr 1974) and untreated human sewage, with a TN/TP of 5.3 (Vollenweider 1968) (Table 4.4).

There are very few studies of waste production from intensive aquaculture in the tropics, primarily since intensive aquaculture is relatively rare (Edwards 1993). Such a database is lacking in the tropics. Costa-Pierce & Roem (1990) studied carp cages in Indonesia (for the larger context of these studies see Costa-Pierce & Soemarwoto 1990). Data are summarised in Table 4.3. This study indicates a substantial difference in waste composition, which probably results from differences in temperatures, feeds, management practices, species and aquatic environments.

The studies of Costa-Pierce & Roem (1990) were modelled on those of Merican & Phillips (1985) for direct comparison of results. In both studies, rates and nutrient quality of sedimenting materials from cages were monitored. Rates of deposition of settleable solids and TP were considerably lower from the tropical cages, although feeding rates and total phosphorus contents of the feeds were similar. Interestingly, the nutrient densities of materials collected from the control stations in the eutrophic, tropical reservoir were higher than the aquaculture wastes sedimenting from cages. Exactly opposite results were found by Merican & Phillips (1985).

Table 4.4   TN/TP ratios for various resources, pollutants and excreta: conversion to molar ratios can be done by multiplying by 2.21 (from Downing & McCauley 1992; see this reference for individual references for each source).

Source	TN/TP ratio
Runoff from unfertilized fields	247.4
Runoff from soils, medium fertility	75.0
Runoff from forested area	71.1
Runoff from rural croplands	60.9
Runoff from fertile soils	33.3
Ground water	28.5
Precipitation	25.4
Runoff from tropical forest	23.5
Runoff from agricultural watershed	20.0
River water	18.9
Mississippi river water	12.2
Seepage from cattle feedlot	8.9
Fertilizer	7.9
Tropical precipitation	7.7
Redfield ratio for phytoplankton	7.2
Feedlot runoff	6.4
Sediments, mesotrophic lake	6.3
Urban stormwater drainage	5.8
Pastureland runoff	4.8
Urban runoff	4.7
Sediments, oligotrophic lake	3.3
Septic tank effluent	2.7
Sediments, eutrophic lake	2.5
Sedimentary rocks	0.8
Felsic rocks	< 0.1
Earth's crust	< 0.1
Mafic rocks	< 0.1

Costa-Pierce & Roem (1990) explained these results by the eutrophic nature of the reservoir, dominated by blooms of *Microcystis* spp., which has a high TP content (Gerloff & Skoog 1954), in comparison with the cages located in oligotrophic coastal waters studied by Merican & Phillips (1985). TN/TP ratios of sedimenting materials from the carp cages were very low, 0.001 (Table 4.3), indicating that wastes were selectively enriched in phosphorus, with trace amounts of TN, possibly due to the high pH in the water, thereby allowing volatilisation of ammonia, and the predominance of algal/bacterial food webs. Bacteria have TN/TP ratios as low as 3.2 (Luria 1960). Unfortunately, the dissolved fraction of the wastes was not quantified in the Costa-Pierce & Roem (1990) study. Although further work is required, the study raises the interesting possibility that if cages are located in eutrophic lakes or reservoirs, and the P content of the feed is low (0.8–1%), that at higher ambient water temperatures wastes from intensive cage aquaculture in the tropics could be of little environmental concern. It was also hypothesised that the high ambient temperatures increased rates of decomposition and nutrient recycling within the cages (Costa-Pierce & Roem 1990).

Careful examination of the nutrient effluent data allows some simple observations:

- System types:
Nutrient flows to the environment from semi-intensive pond aquaculture are small during normal operations in comparison with intensive systems (Schwartz & Boyd 1994). Systems of most environmental concern are intensive cage, floating net pen, raft and other 'open' systems such as raceways. Edwards (1993) calculated that intensive systems at an FCR of 1.5 added 36% more TN and 14% more TP than semi-intensive pond aquaculture systems.
- Waste quality:
The total waste stream from an aquaculture operation consists of settleable and dissolved solids and dissolved, inorganic and organic fractions of TN and TP. Scientific reports have shown a direct relationship between the TP content of added feeds and the TP content of wastes discharged (Solbé 1982; Rosenthal *et al.* 1987; Lall 1991). Intensive aquaculture effluents contain most of the nutrient loads in the dissolved TN and organic TP fractions. The TN/TP ratios of waste solids are highly enriched in TP and are comparable to untreated sewage.
- Waste quantity:
Nutrient loading rates to the environment are related to the scale and intensity of the operation, amount of waste treatment, and level of connection to the environment. Mean discharge rates for each tonne of fish produced are 510 kg of solids, 108 kg of TN and 19 kg of TP (Table 4.3). Composition and impacts of aquaculture wastes are variable, but study results indicate they are pulsed and can be contained with proper engineering. However, results from the temperate zone studies may not be applicable to aquaculture systems in the tropics (Table 4.3).

4.4   COMPARISONS OF AQUACULTURE, AGRICULTURE AND DOMESTIC WASTES

In some nations where aquaculture is not a traditional pursuit, systems are regulated as industrial operations. In these cases, aquaculture is regulated separately from the diversity of systems that comprise modern agriculture. In Minnesota, USA, for example, aquaculture is regulated both as if it were an intensive animal feedlot and an industry. Aquaculture facilities having more than 1000 'animal units' are considered 'industrial', and wastes produced from such facilities are regulated more stringently as 'industrial waste' (Wilcox 1994). If intensive aquaculture can produce 'industrial waste', it is fair to ask, therefore, whether its nutrient contributions and environmental impacts are more problematic than other, more recognisable human impacts.

In the USA, agriculture is by far the largest contributor to nutrient pollution of surface waters (EPA 1978), and in some central states agriculture is also the largest contributor to groundwater pollution (Leete 1991). In the Netherlands, animal feed-lots have produced such water quality and other environmental problems that Armstrong (1988) said the country is 'marooned on a mountain of manure'. In Asia, agriculture, soil erosion and domestic wastes have contributed to the severe degra-dation of water quality of most watercourses in the region.

Examples of phosphorus export coefficients from various land uses in the USA are shown in Fig. 4.1. These values are similar to those reported by Reckhow & Simpson (1980). Figure 4.1 also shows a comparison of TP discharges from these sources with channel catfish ponds (Schwartz & Boyd 1994), salmonid cage culture in the USA (Axler *et al.* 1994), and with intensive cage aquaculture of carp in Indonesian reser-voirs (Costa-Pierce & Roem 1990).

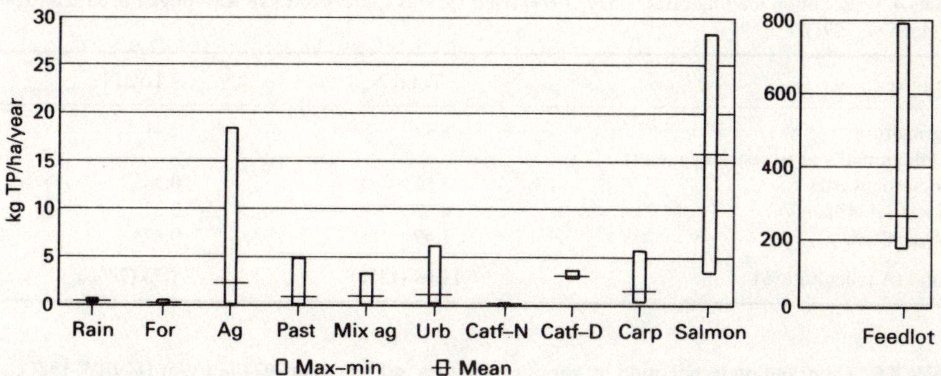

Fig. 4.1 Phosphorus export in kg/ha/year from various land and water uses. Modified from EPA (1980) and Reckhow & Simpson (1980). Rain, precipitation; for, Forestry; Ag, Row crop agriculture; Past, Pasture; Mix ag, Mixed agricultural; Urb, Urban; Catf-N, Catfish ponds during normal operations; Catf-D, Catfish ponds during draining for harvesting; Carp, Carp cage culture in Indonesia (Costa-Pierce & Roem, 1990); Salmon, Salmon cage culture in Minnesota (Axler *et al.* 1994); Feedlot, Feedlot manure storage.

During normal operations of channel catfish ponds, TP releases are comparable to precipitation. However, during harvesting, mean TP discharges are comparable to nutrient discharges from runoff from intensive agriculture. Cage culture of salmonids in Minnesota, USA, produced a mean waste load comparable to the highest TP rates recorded for runoff from intensive agriculture, while carp culture in cages in Indo-nesia produced mean TP loading rates comparable to urban storm-water runoff. These TP export coefficients of aquaculture wastes produced by both semi-intensive and intensive operations show clearly that the magnitude of the waste stream from commercial aquaculture is comparable to many common agriculture and other non-point pollution sources, but is far below the nutrient export of wastes reported for intensive animal feedlots (Fig. 4.1).

In terms of quantity, there have been a number of comparisons of the nutrients from aquaculture effluents with nutrients from other sources. For example, Lake

Kasumigaura in Japan receives effluents from aquaculture operations producing 8400 t of common carp. However, Watanabe (1991) estimated that 87% of the pollution load to the lake was due to agriculture, domestic waste and industrial effluents (Table 4.5). In Finland, commercial aquaculture is a minor source of nutrient pollution. Forestry and heavy industry account for 96% of the solids, 98% of the BOD, 94% of the TP and 87% of the TN loading to Finnish watercourses (Table 4.6). In addition, the pollution potential from domestic wastes world-wide is enormous in scope. About $3.8 \times 10^9 \, \text{m}^3$ of sewage effluent per year is discharged to groundwaters from septic systems in the USA (EPA 1986). This discharge is highly enriched in TN (about 3000 μM TN) and TP (400 μM TP).

Table 4.5 Pollution loading rates (t/day, 1991) from various sources to Lake Kasumigaura, Japan (from Watanabe, 1991).

Polluter	Total N	Total P
Agriculture	5.99	0.31
(all animal and terrestrial sources)		
Domestic wastes	3.56	0.34
Industrial effluents	0.60	0.33
Aquaculture	1.49	0.27
Total (Aquaculture%)	11.64 (13%)	1.25 (22%)

Table 4.6 Contribution to pollution by various industries in Finland in 1978 in t/year (FNBW 1981).

Source	Solids	BOD	TN	TP	TN/TP Ratio
Forestry	87 922	231 020	3 340	505.1	6.6
Industry	10 160	4 382	2 380	73.1	32.6
Agriculture	2 003	4 797	155	30.1	5.1
Aquaculture	1 800	900	240	60.1	4.0
Aquaculture as % of total	2	0.4	4	9	

Aquaculture is a very rare occupation throughout the world. Edwards (1993) states that even though Asia accounts for 85% of total world aquaculture production, 'Aquaculture is far less widespread in Asia than is widely supposed. Perhaps less than 1% of farmers are involved in aquaculture in the region.' Therefore, while the nutrient quality of aquaculture wastes (TN/TP ratios) is comparable to untreated human sewage and urban runoff, the magnitude of total discharge of aquaculture wastes is orders of magnitude lower owing to the fact that aquaculture is an uncommon occupation throughout the world. On a larger scale, even the most intensive aquaculture can produce only localised effects in comparison with modern industry, human wastes and agriculture.

Given these comparisons, it is fair to inquire whether or not the overall regulatory environment concerning aquaculture is evolving in a scientific manner and a supporting manner, and whether aquaculture is in a subservient role to the political process. It is fair to ask whether the regulatory and accompanying political environments are evolving in ways to slow or halt the growth of aquaculture purposely. If government agencies at whatever level are taking roles opposed to aquaculture developments because they oppose the addition of any new sources of nutrients to natural ecosystems, then it is fair to ask regulators to apportion regulatory and monitoring efforts accordingly, and to target the largest polluters, which, according to the previous review, are agriculture, especially feedlots, industry and human wastes (sewage and septic tanks). Given the sorry state of global fisheries it is also fair to ask those politically opposed to aquaculture development whether this opposition is in the public's best interest.

Environmental impacts of aquaculture systems on aquatic ecosystems are related to the location of the systems, intensity of operations, the morphology, physical oceanography/limnology and hydrological nature of the receiving ecosystem, and the trophic status and assimilative characteristics of receiving waters. Choice of proper site locations for intensive aquaculture are critical in avoiding adverse environmental impacts. In this regard, cage aquaculture is the most problematic in terms of direct conflicts with environmental concerns.

The best sites for cage aquaculture are deep, sheltered bays or coves with stable water columns and no deoxygenation events, having low retention times of water (e.g. highly flushed), near to good transportation networks. Sheltered areas of this type are prime sites for other uses and numerous conflicts have developed. In addition, embayments or sills can restrict water flow and currents, leading to accumulation of aquaculture wastes and environmental degradation. Site criteria for many coastal and freshwater areas have been developed (Weston 1986; Lumb 1989), but the integration of social and community concerns with the technical aspects are still lacking in most instances.

4.5   AQUACULTURE NUTRIENTS AND LAKE EUTROPHICATION

There is a large body of scientific work on the ecological status of natural aquatic ecosystems and their responses to nutrient pollution, especially phosphorus (P). Classic studies of pelagic ecosystems have generalised that primary productivity in fresh waters is limited by P (Schindler 1977, 1978) and that marine systems are limited by nitrogen (N) (Ryther & Dunstan 1971; Goldman *et al.* 1973). These studies predicted that increases in nutrient loadings would lead to eutrophication, e.g. that linear increases in the P concentrations of receiving waters would result in linear increases in phytoplankton biomass and productivity, accompanied by changes in the community structure of the phytoplankton towards blue–green algae with low TN/TP ratios

(Horne & Goldman 1994). It has been found that the amount of biologically available phosphorus for algal growth is 10–30% of the TP (Lee *et al.* 1980).

Eutrophication is a natural process of 'ageing' of water bodies due to an increase in nutrient loading, to which humans contribute by accelerating the process, termed 'cultural eutrophication'. As pointed out by Edwards (1993), in fertile water aquaculture systems, such as semi-intensive ponds, the rate of eutrophication is increased by using fertilisers and feeds to increase natural foods and to feed fish maintained at biomasses far exceeding natural conditions. Ryther (1971) termed this 'controlled eutrophication', where fertile water aquaculture attempts to achieve a balance between rates of feed/fertiliser inputs and oxygen produced by primary producers, with the outputs of organisms grown and oxygen consumed in respiration. In contrast, intensive, recirculating aquaculture systems are heterotrophic systems depending on complete dietary and energy rations provided by the culturist and inputs of large amounts of oxygen provided by artificial aeration.

If the waste stream of semi-intensive and intensive aquaculture is comparable in quality (TN/TP ratios) to most agricultural wastes, but differs only in magnitude, a number of empirical models can be used to predict the magnitude of TP in lakes in response to various TP additions. A list of nine such models used for lentic waters and their requirements of input data are given in Table 4.7. In a classical sense, once the ambient TP concentrations have been determined, an inferred lake trophic status is defined (Table 4.8). Furthermore, TP levels have been used along with Secchi disk transparencies and chlorophyll (Chl) levels to define a trophic state index as a guideline for evaluating the changed trophic status of standing water bodies (Table 4.9).

Table 4.7 Empirical models used to predict phosphorus levels in natural waters with various dependent variables.

Authors	Dependent variables
Dillon & Rigler (1974)	Areal phosphorus loading (APL), hydraulic detention time (HDT), phosphorus retention coefficient (PRC)
Vollenweider (1975)	APL, mean depth, HDT
Chapra & Tarapchak (1976)	APL, areal water loading (AWL)
Vollenweider (1976)	APL, AWL, HDT
Jones & Bachmann (1976)	APL, mean depth, HDT
Reckhow (1979)	APL, AWL
Canfield & Bachmann (1981)	APL, HDT, mean depth
Nurnberg (1984)	APL, AWL, PRC, internal P loading, sediment area in contact with anoxic waters, duration of sediment anoxia, rate of P release from anoxic sediments, lake surface area

Table 4.8   Total phosphorus levels used to define trophic status of lakes (from Reckhow & Simpson 1980).

TP Concentration (μg/l)	Trophic state
< 10	Oligotrophic
10–20	Mesotrophic
20–50	Eutrophic
> 50	Hypereutrophic

Table 4.9   Carlson's trophic state index for natural lakes (from Carlson 1977).

TSI < 30	Classic oligotrophy. Clear water, oxygen throughout the year in the hypolimnion. Salmonid fisheries.
TSI 30–40	Deep lakes have classic oligotrophy. Shallow lakes show summer anoxia in hypolimnion.
TSI 40–50	Most lakes show summer anoxia in the hypolimnion.
TSI 50–60	Boundary of classic eutrophic lakes. Decreased transparency, summer anoxia hypolimnion, macrophyte problems, warm water fisheries.
TSI 60–70	Eutrophic lakes. Dominance of blue-green algae, some algal blooms, extensive macrophytes.
TSI 70–80	Hypereutrophic lakes. Production limited by light; persistent algal blooms.
TSI 80–90	Algal scums, fish kills, few macrophages, dominance of carps, catfish, etc.

Trophic state index (TSI) can be calculated from secchi disc (SD) as: $\text{TSI} = 10\,(6\text{–ln SD}/\ln 2)$; from Chl as: $\text{TSI} = 10\,(6\text{–}[2.04\text{–}0.68\ln \text{Chl}/\ln 2])$; from TP as: $10\,(6\text{–}([\ln 48/\text{TP}]/\ln2))$.

Classical empirical models developed by Dillon & Rigler (1974, 1975) have been used to determine the relationship between TP concentrations and algal concentrations (Chl) as a measure of lake trophic status. However, these models are not always applicable to evaluating eutrophication potential from aquaculture since the TP in aquaculture wastes is mostly in the organic fraction, its composition is variable and nutrient loadings are highly 'pulsed'.

One key to evaluating impacts of added nutrients on the eutrophication of pelagic ecosystems is the limiting nutrient concept of Vollenweider (1976). Vollenweider showed a linear, logarithmic relationship between early spring TP concentrations and average summer Chl concentrations in lakes. However, if water bodies are not limited by nutrients but by physical water movements (mixing, dilution) or light, the relationship between TP and Chl falls apart, and the classic biological responses to eutrophication may not occur. Furthermore, recent studies point out that nutrient limitation varies with existing trophic status. TP has been shown to be the best predictor for Chl for waters of low fertility (TP $<$ 200 mg/m^3), while TN and TP predict Chl levels at TP $>$ 200 mg/m^3 (Seip 1994).

Analysis of the TP–Chl relationship has shown that the relationship is not linear but is sigmoid in shape and that TN concentrations have a significant effect on Chl when TP is high (McCauley *et al.* 1989). It appears that both TP and TN affect Chl concentrations, with the magnitude of effects by TN increasing dramatically where TP concentrations are high (Ahlgren 1980). These studies were conducted on natural lakes with varying retention times. Recent review of the limiting nutrient concept has also shown that the classical TP and Chl relationship is influenced by TN concentrations. In a review of 62 lake studies, Elser *et al.* (1990) found that, in most studies, phytoplankton was limited by both TN and TP, and that when these nutrients were added to natural phytoplankton populations (rather than laboratory cultured ones) they stimulated phytoplankton growth over each nutrient added singly.

These are important findings to evaluate nutrient impacts of aquaculture wastes on lentic water bodies in the temperate zone. Since the TN/TP ratio of aquaculture effluents is so low, eutrophication would be expected if aquaculture wastes were added to oligotrophic lakes. In eutrophic to hypereutrophic situations, however, management of aquaculture wastes by controlling the TP levels would have little or no effect on eutrophication, whereas changes in TN loading might cause orders of magnitude reductions in Chl (McCauley *et al.* 1989).

Aquaculture obviously has comparatively greater environmental impacts on vulnerable ecosystems. Unfortunately, largely owing to the 'new' or 'primitive' nature of aquaculture in many countries, enough worst-case scenarios have emerged to pit aquaculture against the environment. There are, for example, intensive aquaculture operations in national parks (see the front cover of Pullin *et al.* 1993), and in wetlands important to natural fisheries that have endangered species of birds (e.g. the Amorient shrimp/prawn facility in Kahuku, Hawaii). Solbé (1982) concluded that TP from fish farming was more of a concern in the oligotrophic waters of the north and west UK and elsewhere in Europe than in eutrophic waters.

There are intensive aquaculture operations located in completely artificial environments (gravel pits, mine pits, irrigation and hydropower reservoirs) (Costa-Pierce & Soemarwoto 1990; Axler *et al.* 1994). Clearly, the largest environmental impact of these developments was the near total destruction of natural aquatic ecosystems by man's machines, and accompanying habitat losses and wholesale landscape degradation. In these artificial lakes, large-scale manipulation of the environment is used to culture fish, and classical limnological relationships between nutrients and the biota are altered. For example, studies of intensive aquaculture operations in artificial lakes where large amounts of aeration and mixing have been used, have found that the TP-Chl relationship is not valid for management decisions because light, not nutrients, becomes the limiting factor to algal photosynthesis.

In studies of cage aquaculture and the aquatic environment of mine pit lakes in Minnesota, USA (Axler *et al.* 1992, 1994; Wilcox 1994), that use extensive mixing and aeration, high TP loading rates have produced eutrophication rates much lower than those predicted by the models of Dillon & Rigler (1974, 1975) and others. Destratification, mixing and aeration of the water column result in a lowering of ambient temperatures and a rapid circulation of algal cells out of the light, decreasing phytoplankton production (Axler *et al.* 1992; D. Wilcox, pers. comm.). Rapid flushing times can also limit the accumulation of a pelagic biomass (Gowen *et al.* 1983).

4.5.1 Aquaculture impacts on lotic systems

For lotic ecosystems, the extent of nutrient pollution from aquaculture may be measured by the ability of a system to absorb pollutants. Engineers have used the term 'assimilative capacity' to refer to the ability of a natural system to assimilate the BOD of oxygen-consuming inputs without lowering natural dissolved oxygen concentrations to levels lower than a 'primary threshold' (Velx 1976). Cairns (1977) defined assimilative capacity in an ecological sense as the 'ability of an aquatic ecosystem to assimilate a substance without degrading or damaging its ecological integrity'. 'Ecological integrity' is defined as the maintenance of ecological structure and function characteristic of that locale (Cairns 1977). Structural integrity is determined from the numbers of organisms and how they are ordered. An abnormal change (increase or decrease) is interpreted as evidence of stress (Patrick 1949).

4.5.2 Aquaculture impacts on benthic ecosystems

Impacts of intensive cage aquaculture on benthic ecosystems have been studied extensively (Enell & Lof 1983; Brown *et al.* 1987; Gowen & Bradbury 1987; Gowen *et al.* 1988; Ritz *et al.* 1989; Gowen & Rosenthal 1993; Johannessen *et al.* 1994). As with the pelagic environment, impacts of sedimenting materials deposited on the benthos are related to the size, intensity and management of farming operations, and to the morphology, bottom topography and physical oceanography/limnology of the site.

In studies from coastal sites receiving wastes from fish cages (cited above) or mussel rafts (Mattsson & Linden 1983), addition of sedimenting materials has been shown to decrease the vertical extent of the top, oxidised (oxygenated) sediment layer, and to increase the extent of anaerobic sediment layer towards the sediment surface. This change lowers sedimentary redox potential. Lower sedimentary redox potentials increase sedimentary oxygen demands (SODs) and stimulate facultative microbial activities.

As a result, microbial metabolisms shift from aerobic to anaerobic, and sulphur and methane reduction occur, increasing oxygen depletion further by increasing sedimentary chemical oxygen demand (COD). Where very high solids and nutrient loading rates occur, microbial processes accelerate to the point where oxygen is depleted, sediments go anaerobic and large amounts of bound, sedimentary phosphorus are released. Bacterial reduction in anaerobic sediments under intensive cage culture operations can be so high as to cause outgassing of methane and hydrogen sulphide (Braaten *et al.* 1983; Samuelsen *et al.* 1988).

In such situations, effects on biological communities can be dramatic. A decrease in species diversity occurs with a simplification of community structure and a large increase in the numbers of a very few opportunistic species (Brown *et al.* 1987; Tsutsumi *et al.* 1991; Johannessen *et al.* 1994). In enclosed basins, the benthos near the cage site can nearly be wiped out owing to deoxygenation.

Summary of the available research on aquaculture's impacts on the benthic environment shows, however, that the pollution processes caused by intensive cage and mussel aquaculture producing detrimental impacts on benthic ecosystems are limited to the immediate vicinity of the aquaculture operations (Table 4.10). While Enell & Lof (1983) measured elevated SODs of 45–55 mg $O_2/m^2/h$ under salmon cages in comparison with a control site of 16 mg $O_2/m^2/h$, these figures are still comparable to the reported ranges of 14–376 mg $O_2/m^2/h$ for natural lakes (Veenstra & Nolen 1991). Flushing, resuspension and dispersion may prevent waste accumulations in many sites (Ackefors 1986; Ackefors & Enell 1990). In addition, long term studies have found that, even for enclosed basins, reported sedimentary oxygen depletions have been brief and could not be attributed solely to the impacts of aquaculture (Gowen & Rosenthal 1993).

Table 4.10 Summary of reports on the extent of environmental impacts by intensive aquaculture on benthic ecosystems.

Study	Impacts
Mattsson & Linden (1983)	Species composition changed from mussel farm up to 20 m away
Brown *et al.* (1987)	Species composition changed from cage edge to 15 m away
Gowen *et al.* (1988)	Species composition changed up to 30–40 m away
Lumb (1989)	Measurable benthic impacts restricted within 50 m of edge of cages
Kupka-Hansen *et al.* (1991)	Species composition changed up to 25 m away
Weston (1991)	Species composition changed up to 90–100 m away
Johannessen *et al.* (1994)	Species composition changed up to 250 m away

There are few studies on the impacts of intensive aquaculture on the benthic ecosystems of tropical lakes and rivers. Whilst it has been assumed that there are fundamental differences in the functioning of temperate and tropical waters (Castagnino 1982), recent reviews have indicated that the majority of tropical lakes and reservoirs are phosphorus-limited and function similarly to temperate zone ecosystems (Salas & Martino 1991).

Deposition of organic materials from aquaculture leaves an organic signature, a 'memory' in the sediments, the extent of which can be read by coring. Organic layers deposited under aerobic conditions are brown, while layers deposited under anaerobic conditions are black. In environments where there is a high degree of soil erosion or natural plankton deposition, these eroded/deposited materials can form a cap over the anaerobic layers since these materials were deposited under highly aerobic conditions (Wilcox 1994). Soil erosion rates of 12 000 kg/ha are common in many agricultural areas (Ellis *et al.* 1978) and where these enter water bodies it can be expected that previous organic layers will be buried. In addition, if a significant amount of fish bioturbation of sediments occurs, mineralisation and aerobic microbial activities increase (Costa-Pierce *et al.* 1983). Therefore, the natural sedimentary profile could recover fairly rapidly if large benthic feeding fish populations exist in the areas of aquaculture development.

While there has been a great deal of work on the impacts of aquaculture on the structural aspects of the environment, fewer studies have been done that address the waste's impacts on functional aspects of the aquatic environment. Assessments of impacts of top-down vs bottom-up functions (Novales-Flamarique *et al.* 1993), impairment of pelagic-benthic coupling mechanisms (Johnson & Wiederholm 1992), or disruption of the chemical ecology mediating inter- and intra-species interactions (Black 1993) have not been completed.

4.6   COMMUNITY-BASED AQUACULTURE SYSTEMS

Development of commercial aquaculture in the West has followed the model of corporate agriculture. In contrast, evolution of aquaculture in the East has long term traditional, cultural, family and community roots. Borgese (1980) stated:

'That aquaculture has a philosophical base in the East and a scientific base in the West has far-reaching implications. In the East, it is culture, it is life: culture to improve life by providing food and employment. It is embedded in the social and economic infrastructure. All that science can and must do is to make this culture more effective. In the West, aquaculture is science and technology, embodied in industry and providing profits: money. It has no social infrastructure. In this, the West has much to learn from the East.'

While aquaculture has cultural roots in the East, it is not a common occupation there (Edwards 1993); however, there are no mysterious roots to its development. Simply stated, population densities in Asia have exceeded the carrying capacities of the natural and agricultural environments in historical times. In order to support these people, new, labour-intensive agriculture innovations such as multiple inter-cropping, multi-storey agriculture and recycling of nutrients through ecologically sophisticated composting and waste recycling regimes became common, but are now endangered. Aquaculture, one of the most complex of all possible agricultural innovations, arose to meet the protein needs of societies that had reached high population densities in historical times. Aquaculture arose as an integral part of a complex polyculture of agricultural systems, and, where it was practised, was intricately entwined into the existing water management system for mixed agriculture.

Aquaculture has traditional, community roots in Asia. In the West, population densities exceeding environmental and agricultural carrying capacities and human needs have only recently reached critical stages, and aquaculture has been adopted. However, the rich West still has nowhere near the pressing food needs of Asia, and generally has numerous and abundant protein choices. Commercial aquaculture in the West has therefore developed on an industrial basis to provide a diversity of products for high-priced, luxury protein markets. In some cases aquaculture operations are considered 'outsiders' by the very communities that host them.

In the last decade there have been radical and reasoned attacks on aquaculture developments from the very communities that should be its natural supporters, namely, rural communities and environmentalists (Barinaga 1990). Western aquaculture has been developed and is perceived by these groups as an industrial rather than a community-based development. While the technical issues of pollution control can be resolved by incorporating ecological or system technologies into aquaculture, public opposition to aquaculture will be a more difficult problem to solve if aquaculture in the West remains an industry producing wastes and requiring evolution of a new regulatory structure (Fig. 4.2).

Clearly, most of the public will not tolerate the addition of any new sources of pollution or the further degradation of the natural environment that is perceived to come at the expense of the degradation of the quality of life. Connections between the disruption of our environment and human health are being made by an increasingly

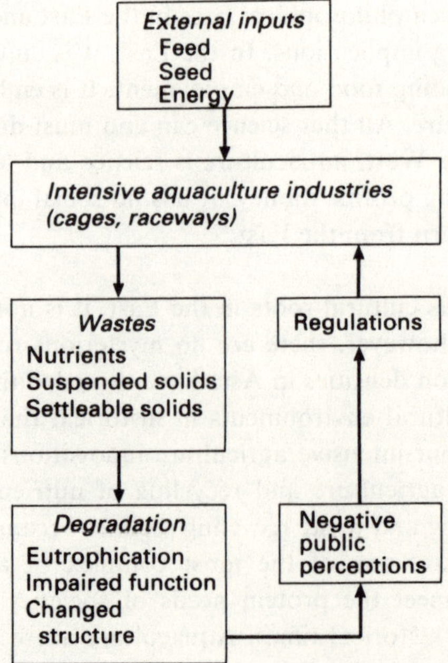

Fig. 4.2 The industrial model of intensive aquaculture produces wastes that need stringent regulations, which create negative public perceptions.

sceptical public determined to fight 'the experts'. In many instances the simple implication of the presence of a chemical implies a hazard and a threat to human health. In order to change the public perception of aquaculture as 'outsiders' or 'industrial polluters', intensive aquaculture operations must plan to become part of a community and a region, and have a wider plan for community development that works with leaders to provide needed inputs, recycle wastes, create a diversity of unprocessed and value-added products, provide local market access, and plan for job creation and environmental enhancement on local and regional scales (Fig. 4.3).

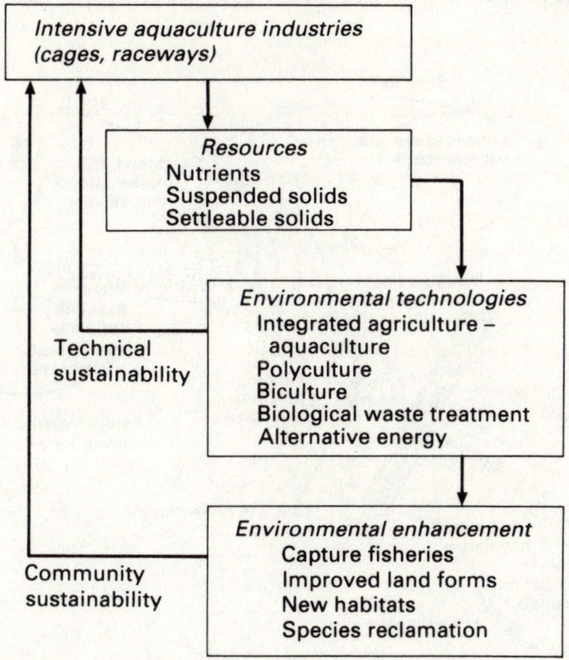

Fig. 4.3 Sustainable, ecological aquaculture uses environmental technologies that turn wastes into resources and opportunities that lead to technical and community sustainability and environmental enhancement.

When viewed from these community development perspectives, intensive aquaculture and the public it intends to serve have many common objectives, especially in isolated rural areas. Examples of such developments, one from the East, in Indonesia, and another from the West, in the Highlands and Islands of Scotland, serve to illustrate the potential for sustainable, community-based, intensive aquaculture development.

In Indonesia, a new reservoir-based cage-culture industry now produces in excess of 10 000 t of common carp per year from 0 t in 1985 (Costa-Pierce & Soemarwoto 1990). The cages have evolved as a new, more efficient and productive part of the existing aquaculture production network for the culture of common carp in West Java (Costa-Pierce 1992). Like all intensive systems, the cages require regular inputs of fish fingerlings and feeds, and produce fresh fish marketed locally or trucked to distant urban markets. The intensive aquaculture has, however, had tremendous spin-off effects for the entire region, in small-scale feed mills, new hatcheries, expansion of rice-fish culture, transportation networks, and other kinds of new and improved employment opportunities (Fig. 4.4). It has been estimated that for every direct job created in the cage-culture developments, 2.5 jobs have been created in the support enterprises (IOE 1992). Costa-Pierce (1995) has termed this example of community aquaculture development 'interactive land/water ecosystem planning'.

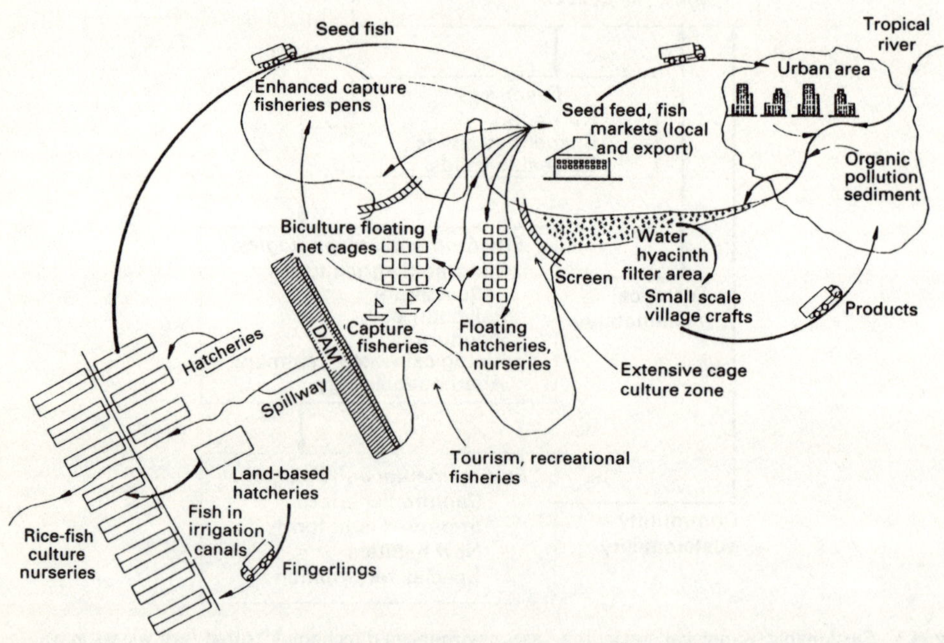

Fig. 4.4 An example of land/water ecosystem planning for the development of reservoir fisheries in Indonesia. The aquaculture production network is highly segmented, with many spin-off and multiplier effects creating new opportunities for employment, environmental rehabilitation and enhancement (from Costa-Pierce 1995).

The community development impact of new aquaculture development in the Highlands and Islands of Scotland has also been impressive (HIDB 1988). In 1993, 320 aquaculture farms in Scotland employed over 1500 persons producing 48 691 t and 21 million smolts (Marine Laboratory Aberdeen 1993). This occurs in a region that in the mid-1960s had lost over half of its 1850s population, and unemployment was about four times the UK national average. Salmon aquaculture in cages has fuelled hope into these fragile and isolated rural communities. Aquaculture provides a significant multiplier effect on the local economies in rural Scotland, increasing both direct and indirect personal spending in small communities.

There is significant on-going evolution of community-based aquaculture in other areas where aquaculture has boomed in the West, such as the Mississippi Delta region of the southern USA, the Snake River Valley, Idaho, USA, the Pacific Coast of Canada, and Norway. Everywhere else, however, aquaculture in the West is very rare, and community support networks have not arisen. The aquaculture production networks (connections between seed, feed and markets) are poorly developed. As a result, while highly advanced technically, many intensive aquaculture operations have few or no community roots, and feeds, seeds, supplies, equipment and markets are procured at great distances from the sites of production. In these circumstances it is easy to see why some community members view aquaculture development as 'all we get is your pollution'.

4.6.1   Aquaculture and resource planning conflicts

Intensive aquaculture systems located in natural environments have received a great amount of attention for their impacts on the environment. Conversely, systems located in artificial environments, such as reservoirs or abandoned gravel or mine pit lakes, would be expected to have few or no impacts on natural ecosystems, unless drinking water issues are involved. After all, these environments are completely unnatural, and so it would be expected that preservationist concerns would be less. While this may be so in many developing countries (Costa-Pierce & Soemarwoto 1990), it is not so in the rich nations where recreation, sport or other conflicts arise, as, for example, in cage aquaculture in abandoned iron mine pits in Minnesota, USA (Axler *et al.* 1994).

Aquaculture is regulated in various ways, on its fish production capacity, the amounts of feeds used, the quality of the effluent, and impacts on receiving waters. For example, in the USA annual aquaculture production per facility of more than 20 000 pounds (9080 kg) requires an NPDEP (National Pollution Discharge Elimination Permit), while Canada requires a Waste Management Permit for cage farms exceeding use of 630 t (dry weight) of feed per year (Caine & Castiedine 1991). In Europe, regulations on emission standards, treatment facilities, feed compositions and amounts, and quality of receiving waters have been adopted (Jensen 1991). This increased concern by the political machinery in many nations is a direct result of the real and potential environmental impacts of intensive aquaculture on natural ecosystems, and due to a lack of planning for proper ecological and community sustainability by aquaculture.

Clearly, the future challenge for intensive aquaculture is to plan for the development of fish production as community development. Success of aquaculture is dependent on its needs for seed, feeds and markets. If aquaculture is planned as grow-out operations only, as feedlots, then the benefits to communities are small. However, if aquaculture is planned as community-based development of a commercial, integrated operation, then hatchery, nursery, feed and marketing facilities need to be created locally, and employment opportunities and the potential for positive community impacts increased. Planning for aquaculture development as community development and environmental enhancement must thereby include planning for aquaculture's vital support industries, and for the use of aquaculture resources and wastes in farming or in environmental enhancement projects. Such a planning effort will have much higher positive impacts on jobs and the environment, and will eventually dissolve the opposition from communities, which will see the newcomers as one of their own.

4.7   DEVELOPMENT OF SUSTAINABLE, ECOLOGICAL AND AQUACULTURE SYSTEMS

Corporate agriculture is the largest source of pollution in the over-industrialised countries. Dent & Anderson (1971) stated that agriculture in these countries has 'ignored the larger ecological framework in which farming is conducted and as a result agricultural production has often exploited the natural environment'. Edwards (1993) states that:

'the problem for developing countries is essentially how to stimulate agricultural (and aquaculture) productivity and profitability without further environmental degradation, in contrast to the need to reduce the level of intensification to a sustainable level in the developed world.'

The Brundtland Commission (WCED 1987) has defined sustainable development as 'the ability to meet the needs of the present without compromising the ability of the future generations to meet their own needs.' Pollution is defined as 'an undesirable change in the physical, chemical or biological characteristics of air, water or land that may or will be harmful to humans and other living organisms' (Odum 1975). Intensive aquaculture operations are feedlots (Wohlfarth & Schroeder 1979) that are energy intensive (Weatherly & Cogger 1977) and protein intensive to the point of being a net consumer, rather than a producer, of fish protein (Schroeder 1980). Intensive aquaculture produces nutrient pollution loads comparable to human sewage (Bergheim & Sivertsen 1981; Bergheim *et al.* 1982). Liao & Mayo (1972) estimated that waste production from salmonid culture in the USA was equivalent to that resulting from a population of 2 million persons. These data, however, must be weighed with the fact that intensive aquaculture is still a minor contributor to total aquatic pollution worldwide.

As currently practised, however, intensive aquaculture is still a contributor to nutrient pollution, no matter how small, and in many countries is the most rapidly growing form of agriculture. Aquaculture is a new pollution source, adding nutrients to already degraded aquatic environments, and, no matter how localised, the concentration of its discharges is comparable to intensive agriculture and untreated human sewage. It is questionable whether the manner in which intensive aquaculture is practised is sustainable.

4.7.1 Low-input sustainable aquaculture systems

As an infant enterprise the world over, aquaculture can ill afford to recreate the sorry history of commercial agriculture, where huge toxic, nutrient and chemical loadings were (and still are) washed down a primitive path of 'the solution to pollution is dilution'. In this regard, modern aquaculture should throw away the model of corporate agriculture and adopt a new strategy, a community-based, low-input, sustainable aquaculture (LISA) model that produces certified organic produce, similar to a LISA strategy now being promoted in agriculture and industry called 'input management' (Odum 1989).

Given this new evolution in LISA, it is opportune to reflect on what are the characteristics of sustainable, ecologically-integrated aquaculture systems, how they function, and to discuss briefly the current evolution of land- and water-based alternatives to the current industrial models. However, while the technology for creating sustainability is vitally important, it is also essential to consider the social ecology of these new developments from the outset in order to articulate what are the most important development goals for sustainable aquaculture systems, or to define what will be required socially in order to develop a systems ecology thinking applied to aquaculture. Folke & Kautsky (1991) concluded that 'one must expand the boundaries and one's actions far beyond the cultivation site, and realise that there is an unavoidable complementarity between the life-support environment and aquaculture production'.

4.7.2 Ecological aquaculture

Sustainable, ecological aquaculture is the development of aquatic farming systems that preserve and enhance the form and functions of the natural environments in which they are situated. It incorporates a shift of the mode of inputs from energy and material subsidies to photosynthetic, waste reuse and natural energy sources to produce net protein gains to society without degradation of natural ecosystems. It is community-based, having positive societal impacts, not just economic benefits. Ecological aquaculture systems are planned using systems ecology approaches that develop aquaculture production networks for various species in a highly diversified, segmented manner, with numerous interconnections that supply inputs and outputs using local resources and recycled wastes and materials, planning for maximal job creation, and closing leaky loops of energy and materials that can potentially degrade natural ecosystems.

Aquaculture depends upon inputs from various food, processing and transportation industries and produces valuable waste waters, manures and fish wastes, all of which are part of an ecological system that can be planned and organised for ecosystem rehabilitation, reclamation and enhancement, not degradation. Ecological aquaculture treats and recycles its own wastes rather than relying upon public subsidies, and integrates people with technologies in new synergies to create new employment and biotechnical advances with earth-centred technologies (wind, solar and biomass). Ecological aquaculture takes a global view, integrating ecological science and sharing technological information with innovation in the global marketplace, avoiding the proprietary.

Some of the characteristics of ecological aquaculture are as follows. It:

(1) preserves the form and functions of natural ecosystems;
(2) derives most of its energy from renewable sources (solar, wind, water, biomass);
(3) is a net protein producer, relying on waste animal- or plant-based proteins for feeds;
(4) does not produce nutrient or chemical pollution;
(5) develops a systems approach to nutrient recycling and regeneration;
(6) plans for ecosystem rehabilitation and enhancement;
(7) is integrated with agriculture;
(8) does not use chemicals or antibiotics harmful to human or ecosystem health;
(9) uses native or resident species;
(10) is integrated with communities to maximise job creation in local industries;
(11) develops enhanced fisheries; and
(12) is a global partner, producing information for the world.

Transitions to a sustainable aquaculture will require a movement from the sewage treatment and assimilative concepts of input and waste management, towards the concepts of ecological waste treatment technologies and input management.

There are very few models of sustainable, ecological aquaculture systems for the world to adopt, but the possibilities for engineering ecological systems that include aquaculture are many (Mitsch & Jorgensen 1989). The model of the Chinese polyculture farm ponds intimately integrated into the agriculture and industrial fabric of southern China serves as one culturally significant example of successful sustainable aquaculture (Li 1987; Ruddle & Zhong 1988), but it is now endangered by China's headlong rush for economic expansion (R. Zweig, pers. comm.). Aquaculture is being integrated into agriculture irrigation schemes in the US south-west (Olsen & Fitz-simmons 1994), and integrated marine pond systems have been developed where shrimp pond effluents are diverted into oyster and mussel rearing ponds (Wang 1990).

For freshwater cage, tank and raceway aquaculture systems, input management, movement from monoculture to polyculture and biculture systems (e.g. one species being grown in a culture system receiving formulated feeds placed above or at higher elevation from a second system below it holding species that are unfed), 'janitorial' polyculture systems (carp in prawn ponds; cleaner fish in salmon cages), and collection of wastes in settling ponds or cage collectors ('diapers') for use in artificial wetlands, show great potential (Fig. 4.5) (Costa-Pierce & Hadikusumah 1990; Bergheim *et al.* 1991; Axler *et al.* 1994; D. Noble, pers. comm.). Data on the operation of cage collectors (Fig. 4.6) show that about 80% of the solids from intensive salmonid cage aquaculture in Minnesota, USA, can be collected (Fig. 4.7). This solid mass contains approximately 65% of the phosphorus added (Fig. 4.8), and is proving to be a very effective growth promoter to plants in an artificial wetland (R. Axler, personal communication).

In Idaho, USA, trout farms produce about 17 million lb (7700 t) of settleable solids. In 1981, a project showed that 2–3 lb (0.91–1.36 kg) of fish manure equalled 1 lb (0.4 kg) of nitrogen in cornfields. The bottom line is that Idaho's trout farms produce enough fertiliser for about 1000 acres of croplands (Yarris 1981). Studies have shown that freshwater aquaculture can cause the increased growth rate of natural fish species, in part because of the natural ingestion of uneaten feeds and feces (Kilambi *et al.* 1976; IOA *et al.* 1990). In a survey of 17 aquaculture farms in Europe, capture fisheries were found to be little affected by fish farm effluents, and in one case there was an improvement in capture fisheries after aquaculture was initiated (Alabaster 1982). For marine cage systems, Folke & Kautsky (1989) proposed to integrate mussel and cage aquaculture. The mussels eat uneaten feeds from cage culture as well as microorganisms filtered.

The world will shortly witness unheard-of advances in solar energy. Associated Press (1994) reported that a breakthrough in solar cell technologies had been achieved which promises to cut the cost of solar energy by 80% in 5–10 years, making it cheaper than coal-fuelled power sources of energy. Intensive, ecological aquaculture systems that incorporate technological advances in biological waste treatment, solar and wind power and integrate these with agriculture can lead the future development of a sustainable aquaculture in the West.

4.8   CONCLUSIONS

Aquaculture development has been slowed, but industry, agriculture and urbanisation are speeding apace. Aquaculture must be judged on its own merits as an intensive farming enterprise and not part of industry. Semi-intensive pond aquaculture produces little nutrient pollution on a routine basis, and the nutrients discharged at harvest can be captured effectively using simple settling ponds. This type of aquaculture should not be regulated by discharge permits if simple treatment methods are incorporated. However, modern, intensive cage, raft and raceway aquaculture is a net energy and protein loss to society and produces pollution comparable to sewage which, in the vast majority of instances, is dumped into receiving waters with little or no treatment.

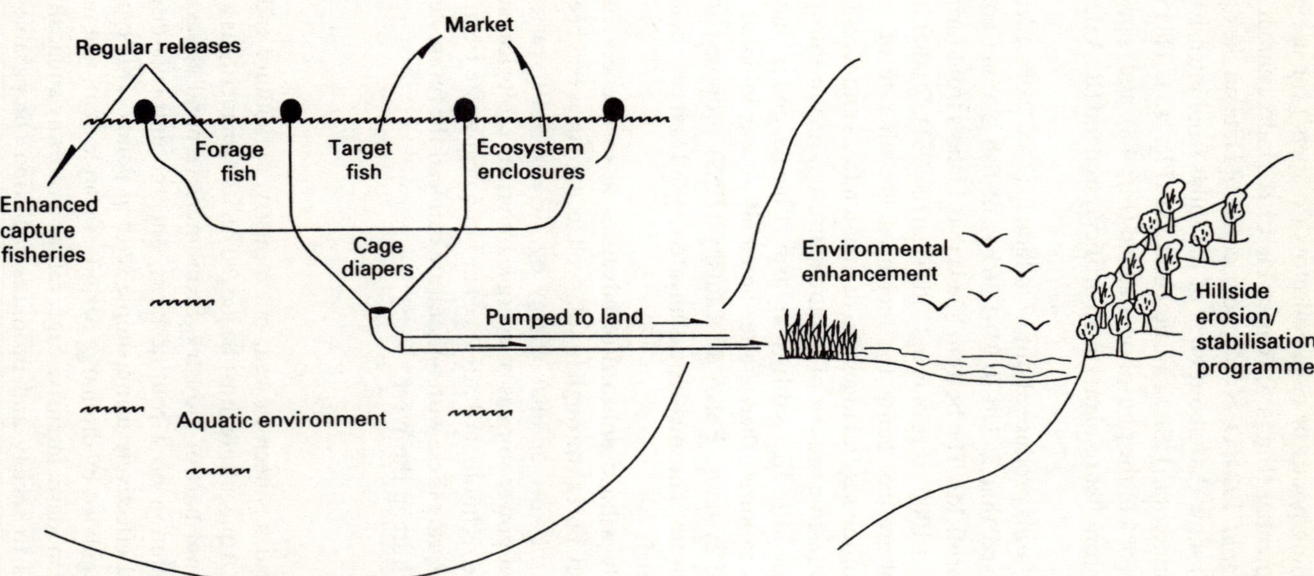

Fig. 4.5   A sustainable model of water-based aquaculture proposed for cage aquaculture of salmonids in mine pit lakes in Minnesota, USA. Wastes are collected and reused for creating artificial wetlands and rehabilitating eroded hillsides. Ecosystem enclosures are used to capture fine and dissolved organic TP escaping from the sides of growout cages and to grow unfed baitfishes, which are released after growout to enhance capture and recreational fisheries.

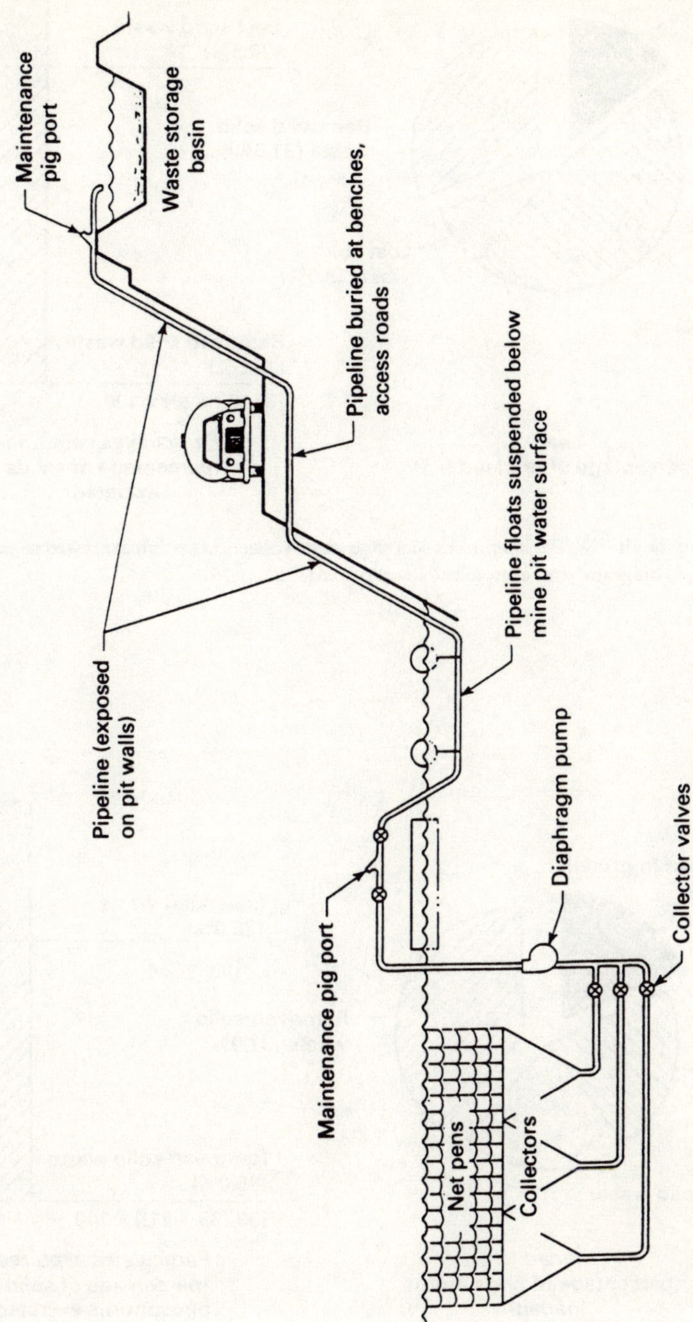

Fig. 4.6 Diagram of the waste collection system set under net pens in a mine pit lake in Minnesota, USA.

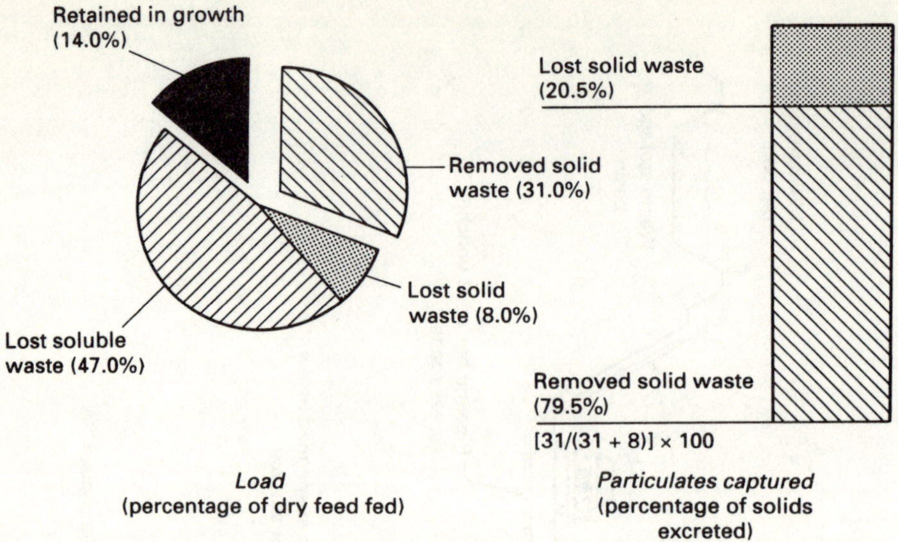

Fig. 4.7 Results of monitoring the operations of a cage waste collection system expressed as percentages of dry feed fed (top pie diagram) and excreted solids (bottom).

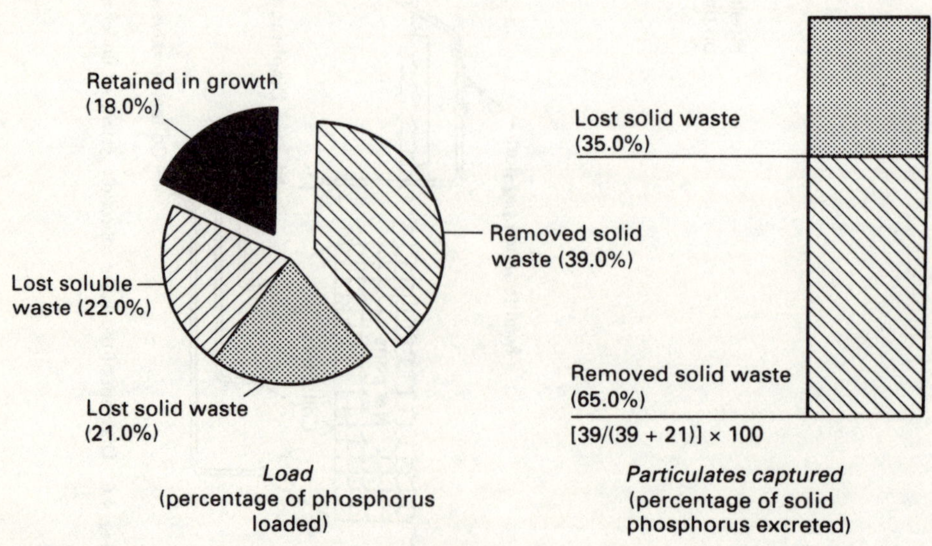

Fig. 4.8 Phosphorus removal by the waste collection system as percentage of feed phosphorus added (top pie diagram) and of solid phosphorous excreted (bottom).

An alternative model of ecological aquaculture development is needed that brings sustainability, ecological methods and systems thinking to intensive aquaculture in a holistic sense incorporating social, economic and wider environmental contexts. These methods will create new opportunities for a wider group of professionals and entrepreneurs to become involved in aquaculture, since new advances will be needed in treatment technology, production management, feed technology, energy technologies, and information, community facilitation and interactions.

REFERENCES

Ackefors, H. (1986) The impact on the environment by cage fish farming in open water. *Journal of Aquaculture in the Tropics*, **1**, 25–33.

Ackefors, H. & Enell, M. (1990) Discharge of nutrients from Swedish fish farming to adjacent sea areas. *Ambio*, **119**, 28–35.

Ahlgren, I. (1980) A dilution model applied to a system of shallow eutrophic lakes after diversion of sewage effluents. *Archivs für Hydrobiologie*, **89**, 17–32.

Alabaster, J. (1982) Survey of fish-farm effluents in some EIFAC countries. In: *Report of the EIFAC Workshop on Fish-Farm Effluents* (ed. J. Alabaster), pp. 5–20. FAO, Rome.

Armstrong, S. (1988) Marooned in a mountain of manure. *New Scientist, 26 November* 1988, 1640, 51–5.

Associated Press (1994) Scientists find cheap way to make solar electricity. *Associated Press Bulletin*, 27 May, 1994, New York.

Axler, R., Larsen, C., Tikkanen, C., McDonald, M & Host, G. (1992) Limnological assessment of mine pit lakes for aquaculture use. *Technical Report NRRI/TR-92/03*. Natural Resources Research Institute, University of Minnesota, Duluth, MN.

Axler, R., Owen, C., Ameel, J., Ruzycki, E. & Henneck, J. (1994) Water quality issues associated with aquaculture: a case study in Minnesota mine pit lakes. *Lake and Reservoir Management*, **9**, 53.

Barinaga, M. (1990) Fish, money, and science in Puget Sound. *Science*, **247**, 631.

Bergheim, A. & Sivertsen, A. (1981) Oxygen consuming properties of effluents from fish farms. *Aquaculture*, **22**, 185–7.

Bergheim, A., Silvertsen, A. & Selmer-Olsen, A. (1982) Estimated pollution loadings from Norwegian fish farms. I. Investigations 1978–1979. *Aquaculture*, **28**, 347–61.

Bergheim, A., Aabel, J. & Seymour, E. (1991) Past and present approaches to aquaculture waste management in Norwegian net pen culture operations. In: *Nutritional Strategies and Aquaculture Waste* (ed. C.B. Cowey & C.Y. Cho), pp. 117–36. University of Guelph, Ontario.

Beveridge, M. (1987) *Cage Aquaculture*. Fishing News Books, Farnham, England.

Black, A. (1993) Predator-induced phenotypic plasticity in *Daphnia pulex:* life history and morphological responses in *Notonecta* and *Chaoborus*. *Limnology and Oceanography*, **38**, 986–96.

Borgese, E.M. (1980) *Seafarm, The Story of Aquaculture*. H.N. Abrams, New York.

Boyd, C.E. (1978) Effluents from catfish ponds during fish harvest. *Journal of Environmental Quality*, **7**, 59–62.

Brown, J., Gowen, R. & McLusky, D. (1987) The effect of salmon farming on the benthos of a Scottish sea loch. *Journal of Experimental Marine Biology and Ecology*, **109**, 39–51.

Braaten, B., Aure, J., Ervik, A. & Boge, E. (1983) Pollution problems in Norwegian fish farming. ICES CM 1983/F: 26.

Caine, G. & Castiedine, A. (1991) Environmental regulations framed to minimize waste in aquaculture effluents in British Columbia. In: *Nutritional Strategies and Aquaculture Wastes* (ed C.B. Cowey & C. Cho), pp. 263–70. University of Guelph, Ontario.

Cairns, J. (1977) Aquatic ecosystem assimilative capacity. *Fisheries*, **2**, 5–7, 24.

Canfield, D. & Bachmann, R. (1981) Prediction of total phosphorus concentrations, chlorophyll *a*, and secchi depths in natural and artificial lakes. *Canadian Journal of Fisheries and Aquatic Science*, **38**, 414–23.

Carlson, R.E. (1977) A trophic state index for lakes. *Limnology and Oceanography*, **22**, 361–9.

Castagnino, W. (1982) *Research on Simple Eutrophication Models for Tropical Lakes*. Pan American Health Organization, Pan American Center for Sanitary Engineering and Environmental Sciences, Lima, Peru (in Spanish).

Chapra, S. & Tarapchak, S. (1976) A chlorophyll *a* model and its relationship to phosphorus loading plots for lakes. *Water Resources Research*, **12**, 1260–4.

Costa-Pierce, B.A. (1992) Rice-fish systems as intensive nurseries. In: *Rice-Fish Research and Development in Asia* (eds C. dela Cruz, C. Lightfoot, B. Costa-Pierce, V. Carangal & M. Bimbao), pp. 117–30. ICLARM, Manila, Philippines.

Costa-Pierce, B.A. (1995) *Roles of Reservoir Fisheries in Interactive, Land/Water Ecosystem Planning for Resettlement*. The World Bank, Washington, DC.

Costa-Pierce, B.A. & Hadikusumah, H. (1990) Research on cage aquaculture systems in the Saguling Reservoir, West Java, Indonesia. In: *Reservoir Fisheries and Aquaculture Development for Resettlement in Indonesia* (eds B.A. Costa-Pierce & O. Soemarwoto), pp. 112–217. ICLARM, Manila, Philippines.

Costa-Pierce, B.A. & Roem, C.M. (1990) Waste production and efficiency of feed use in floating net cages in a eutrophic tropical reservoir. In: *Reservoir Fisheries and Aquaculture Development for Resettlement in Indonesia* (eds B.A. Costa-Pierce & O. Soemarwoto), pp. 257–71. ICLARM, Manila, Philippines.

Costa-Pierce, B.A. & Soemarwoto, O. (eds) (1990) *Reservoir Fisheries and Aquaculture Development for Resettlement in Indonesia*. ICLARM, Manila, Philippines.

Costa-Pierce, B.A., Malecha, S.R., Clay, L. & Laws, E.A. (1983) Benthic microbial biomass, organic matter, and respiration changes in prawn (*Macrobrachium rosenbergii*) ponds with and without fish polycultures. In: *Proceedings of the First International Conference on Warmwater Aquaculture – Crustacea* (eds G. Rogers, R. Day & A Lim), pp. 420–8. Brigham Young University, Laie, Hawaii.

Cowey, C.B. & Cho, C.Y. (eds) (1991) *Nutritional Strategies and Aquaculture Waste*. University of Guelph, Ontario.

Davlin, A. (1991) The 90s: A booming decade for the aquaculture industry! *Analyst's Reports*, **3**, 1–6.

Dent, J.B. & Anderson, J. (1971) Systems, management and agriculture. In: *Systems Analysis in Agricultural Management* (eds J. Dent & J. Anderson), pp. 3–14. Wiley and Sons, New York.

DePauw, N. & Joyce, J. (eds) (1991) *Aquaculture and Environment. Special Publication*, **16**. European Aquaculture Society, Ghent, Belgium.

Dillon, P. & Rigler, F. (1974) The phosphorus-chlorophyll relationship in lakes. *Limnology and Oceanography*, **19**, 767–73.

Dillon, P. & Rigler, F. (1975) A simple method for predicting the capacity of a lake for development based on lake trophic status. *Journal of the Fisheries Research Board of Canada*, **32**, 1519–31.

Downing, J.A. & McCauley, E. (1992) The nitrogen:phosphorus relationship in lakes. *Limnology and Oceanography*, **37**, 936–45.

Edwards, P. (1993) Environmental issues in integrated agriculture–aquaculture and wastewater-fed fish culture systems. In: *Environment and Aquaculture in Developing Countries* (ed. R. Pullin, H. Rosenthal & J. Maclean), pp. 139–70. *ICLARM, Manila, Philippines*.

Ellis, J., Tackett, D. & Carter, R. (1978) Discharge of solids from fish ponds. *Progressive Fish-Culturist*, **40**, 165–6.

Elser, J., Marzolf, E. & Goldman, C. (1990) Phosphorus and nitrogen limitation of phytoplankton in the freshwaters of North America: a review and critique of experimental enrichments. *Canadian Journal of Fisheries Aquatic Sciences*, **47**, 1468–77.

Enell, M. & Lof, J. (1983) Environmental impact of aquaculture: sediment and nutrient loadings from fish cage culture farming. *Vatten*, **39**, 364–75.

EPA (1978) National water quality inventory. Report *EPA-440/4-78-001*. Office of Water Regulations and Standards, US, EPA, Washington, DC.

EPA (1980) Modelling phosphorus loading and lake response under uncertainty: a manual and compilation of export coefficients. *Report EPA 440/5-80-011*. Office of Water Regulations and Standards, US EPA, Washington DC.

EPA (1986) *Septic Systems and Groundwater Protection: A Program Manager's Guide and Reference Book*. Office of Groundwater Protection, US EPA, Washington, DC.

FNBW (Finland National Board of Waters) (1981) Industrial water statistics in 1977–1978. National Board of Water, Helsinki, Finland (in Finnish with English Summary).

Folke, C. & Kautsky, N. (1989) The role of ecosystems for a sustainable development of aquaculture. *Ambio*, **18**, 234–43.

Folke, C. & Kautsky, N. (1991) Ecological economic principles for aquaculture development. In: *Nutritional Strategies and Aquaculture Wastes* (eds C.B. Cowey & C.Y. Cho), pp. 207–22. University of Guelph, Ontario.

Gerloff, G. & Skoog, F. (1954) Cell contents of nitrogen and phosphorus as a measure of the availability for growth of *Microcystis aeruginosa*. *Ecology*, **35**, 348–53.

Goldman, J., Tenore, K. & Stanley, H. (1973) Inorganic nitrogen removal from wastewater: effect on phytoplankton growth in coastal marine waters. *Science*, **180**, 955–6.

Gowen, R. & Bradbury, N. (1987) The ecological impact of salmonid farming in coastal waters: a review. *Oceanography and Marine Biology, an Annual Review*, **25**, 563–75.

Gowen, R. & Rosenthal, H. (1993) The environmental consequences of intensive coastal aquaculture in developed countries. In: *Environment and Aquaculture in Developing Countries* (eds R.S.V. Pullin, H. Rosenthal & J. Maclean), pp. 102–15. ICLARM, Manila, Philippines.

Gowen, R., Tett, P. & Jones, K. (1983) The hydrography and phytoplankton ecology of Loch Ardbhair: a small sea loch on the west coast of Scotland. *Journal of Experimental Marine Biology and Ecology*, **71**, 1–16.

Gowen, R., Brown, J., Bradbury, N. & McLusky, D. (1988) *Investigations into Benthic Enrichment, Hypernutrification and Eutrophication Associated with Mariculture in Scottish Coastal Waters (1984–1988)*. Department of Biology, University of Stirling, Scotland.

HIDB (Highlands and Islands Development Board) (1988) The place of fish farming in the Highlands and Islands of Scotland. Paper presented at the International Fish Farming Conference, Inverness, Scotland (mimeo).

Horne, A. & Goldman, C. (1994) *Limnology*, 2nd edition. McGraw Hill, New York.

ICES (International Council for the Exploration of the Sea) (1989) Report of the working group on the environmental impact of mariculture. ICES C.M. 1989/F:11.

IOA (Institute of Aquaculture), Institute of Freshwater Biology and Institute of Terrestrial Ecology (1990) *Fish Farming and the Scottish Freshwater Environment*. Nature Conservancy Council, Edinburgh.

IOE (Institute of Ecology) (1992) *Socioeconomic Monitoring Impact Study of the Cirata I Project*. IOE, Padjadjaran University, Bandung, Indonesia.

Iwama, G. (1991) Interactions between aquaculture and the environment. *Critical Reviews in Environmental Control*, **21**, 177–216.

Jensen, J. (1991) Environmental regulation of freshwater fish farms in Denmark. In: *Nutritional Strategies and Aquaculture Waste* (eds C.B. Cowey & C.Y. Cho), pp. 251–62. University of Guelph, Ontario.

Johannessen, P., Botnen, H. & Tvedten, O. (1994) Macrobenthos: before, during and after a fish farm. *Aquaculture and Fisheries Management*, **25**, 55–66.

Johnson, R.K. & Wiederholm, T. (1992) Pelagic-benthic coupling – the importance of diatom interannual variability for population oscillations of *Monoporeia affinis*. *Limnology and Oceanography*, **37**, 1596–607.

Jones, J. & Bachmann, R. (1976) Prediction of phosphorus and chlorophyll levels in lakes. *Journal of the Water Pollution Control Federation*, **48**, 2176–82.

Kilambi, R., Hoffman, C., Brown, A., Adams, J. & Wickizer, W. (1976) *Effects of Cage Culture Fish Production upon the Biotic and Abiotic Environment of Crystal Lake, Arkansas*. Department of Zoology, University of Arkansas, Fayetteville, Arkansas.

Kupka-Hansen, P., Pittman, K. & Ervik, A. (1991) Organic waste from marine fish farms – effects on the seabed. In: *Marine Aquaculture and Environment*, pp. 105–19. Nordic Council of Ministers, Copenhagen.

Lall, S. (1991) Digestibility, metabolism and excretion of dietary phosphorus in fish. In: *Nutritional Strategies and Aquaculture Waste* (eds C.B. Cowey & C.Y. Cho), pp. 21–36. University of Guelph, Ontario.

Lee, G., Jones, R. & Rast, W. (1980) Availability of phosphorus to phytoplankton and its implications for phosphorus management strategies. In: *Phosphorus Management Strategies for Lakes* (eds R. Loehr,

C. Martin & W. Rast), pp. 259–308. Ann Arbor Science Publications, Ann Arbor, MI.

Leete, J. (1991) Ground water quality and management in Minnesota. *Journal of the Minnesota Academy of Science*, **56**, 34–43.

Li, S. (1987) Energy structure and efficiency of a typical Chinese integrated fish farm. *Aquaculture*, **65**, 105–18.

Liao, P. & Mayo, R. (1972) Salmonid hatchery reuse systems. *Aquaculture*, **1**, 317–35.

Loehr, R.C. (1974) Characteristics and comparative magnitude of non-point sources. *Journal of the Water Pollution Control Federation*, **46**, 1849–72.

Lumb, C. (1989) Self pollution by Scottish salmon farms? *Marine Pollution Bulletin*, **20**, 375–9.

Luria, S. (1960) The bacterial protoplasm: composition and organization. In: *The Bacteria: A Treatise on Structure and Function* (eds I. Gunsalers & R. Stanier), pp. 1–34. Academic Press, New York.

McCauley, E., Downing, J. & Watson, S. (1989) Sigmoid relationships between nutrients and chlorophyll among lakes. *Canadian Journal of Fisheries and Aquatic Sciences*, **46**, 1171–5.

Mäkinen, T. (1991) Nutrient load from marine aquaculture. In: *Marine Aquaculture and Environment* (ed T. Mäkinen), pp. 1–8. Nordic Council of Ministers, Copenhagen, Denmark.

Marine Laboratory, Aberdeen (1993) *Scottish Fish Farms. Annual Production Survey, 1993*. The Scottish Office, Department of Agriculture and Fisheries for Scotland, Aberdeen, Scotland.

Mattson, J. & Linden, O. (1983) Benthic macrofauna succession under mussels. *Mytilus edulis*, cultured on hanging long lines. *Sarsia*, **68**, 97–102.

Merican, Z. & Phillips, M. (1985) Soid waste production from rainbow trout. *Salmo gairdneri* Richardson, cage culture. *Aquaculture and Fisheries Management*, **1**, 55–69.

Mitsch, W. & Jorgensen, S. (eds) (1989) *Ecological Engineering: An Introduction to Ecotechnology*. Wiley, New York.

Novales-Flamarique, I., Griesbach, S., Parent, M., Cattaneo, A. & Peters, R. (1993) Fish foraging behavior changes plankton-nutrient relations in laboratory microcosms. *Limnology and Oceanography*, **38**, 290–8.

Nurnberg, G. (1984) The prediction of internal phosphorus load in lakes with anoxia hypolimnia. *Limnology and Oceanography*, **29**, 111–24.

Odum, E.P. (1975) *Ecology*. Holt-Sanders, New York.

Odum, E.P. (1989) Input management production systems. *Science*, **243**, 177–81.

Olsen, M. & Fitzsimmons, K. (1994) Integration of catfish and tilapia production with irrigation of cotton. In: *Book of Abstracts*, p. 172. World Aquaculture '94, New Orleans, LA.

Patrick, R. (1949) A proposed biological measure of stream conditions based on a survey of the Conestoga Basin, Lancaster County, Pennsylvania. *Proceedings of the Academy of Natural Science Philadelphia*, **101**, 277–341.

Phillips, M. (1985) *The Environmental Impact of Cage Culture on Scottish Freshwater Lochs. Report to the Highlands and Islands Development Board*. Institute of Aquaculture, University of Stirling, Scotland.

Pullin, R.S.V., Rosenthal, H. & Maclean, J.L. (eds) (1993) *Environment and Aquaculture in Developing Countries*. ICLARM, Manila, Philippines.

Reckhow, K. (1979) Quantitative techniques for the assessment of lake quality. *Report EPA 440/5-79-015*. US EPA, Washington, DC.

Reckhow, K. & Simpson, J. (1980) A procedure using modelling and error analysis for the prediction of lake phosphorus concentration from land use information. *Canadian Journal of Fisheries and Aquatic Sciences*, **37**, 1439–48.

Redfield, A. (1958) The biological control of chemical factors in the environment. *American Scientist*, **46**, 205–21.

Ritz, D., Lewis, M. & Shen, M. (1989) Response to organic enrichment of infaunal macrobenthic communities under salmonid sea cages. *Marine Biology*, **103**, 211–14.

Rosenthal, H., Weston, D., Gowen, R. & Black, E. (1987) Environmental impacts of mariculture. *ICES Cooperative Research Report*, **154**. Copenhagen, Denmark.

Ruddle, K. & Zhong, G. (1988) *Integrated Agriculture–Aquaculture in South China: the Dyke–Pond System of the Zhujiang Delta*. Cambridge University Press, Cambridge.

Ryther, J. (1971) Recycling human waste to enhance food production from the sea. *Environmental Letters*, **1**, 79–87.

Ryther, J. & Dunstan, W. (1971) Nitrogen, phosphorus, and eutrophication in the coastal marine environment. *Science*, **171**, 1008–13.

Salas, H. & Martino, P. (1991) A simplified phosphorus trophic state model for warm-water tropical lakes. *Water Research*, **25**, 341–50.

Samuelsen, O., Ervik, A. & Solheim, E. (1988) A qualitative and quantitative analysis of the sediment gas and diethylether extract of the sediment from salmon farms. *Aquaculture*, **74**, 277–85.

Schindler, D. (1977) Evolution of phosphorus limitation in lakes. *Science*, **195**, 260–62.

Schindler, D. (1978) Factors regulating phytoplankton production and standing crop in the world's freshwaters. *Limnology and Oceanography*, **23**, 478–86.

Schroeder, G. (1980) Fish farming in manure loaded ponds. In: *Integrated Agriculture–Aquaculture Farming Systems* (eds R.S.V. Pullin & Z. Shehadeh), pp. 73–86. ICLARM, Manila, Philippines.

Schwartz, M. & Boyd, C.E. (1994) Effluent quality during harvest of channel catfish from watershed ponds. *Progressive Fish-Culturist*, **56**, 25–32.

Seip, K.L. (1994) Phosphorus and nitrogen limitation of algal biomass across trophic gradients. *Aquatic Science*, **56**, 16–28.

Solbé, J. (1982) Fish farm effluents: a United Kingdom survey. *EIFAC Technical Paper*, **41**, 29–56. FAO, Rome.

Solbé, J. (1987) Effluent control and the UK fish farmer. In: *The English Fish Farming Conference*, pp. 77–85. Hampshire College of Agriculture, 12–13 September 1987.

Tsutsumi, H., Kikuchi, T., Tanaka, M., Higashi, T., Imasaka, K. & Miyazaki, M. (1991) Benthic faunal succession in a cove organically polluted by fish farming. *Marine Pollution Bulletin*, **23**, 233–8.

UMA (UMA Engineering, Inc.) (1988) Waste water treatment in aquaculture facilities. Report to the Province of Ontario, Canada. Mississauga, Ontario.

Velx, C. (1976) Stream analysis–forecasting waste assimilative capacity. In: *Water Resources and Pollution Control* (eds H. Gehm & J. Bregman), p-p. 216–61. Van Nostrand Reinhold, New York.

Veenstra, J. & Nolen, S. (1991) In-situ sediment oxygen demand in five southwestern US lakes. *Water Research*, **25**, 351–4.

Vollenweider, R.A. (1968) Water management research. *Tech. Report DAS/CSI/68.27*. OECD, Paris, France.

Vollenweider, R.A. (1975) Input–output models with reference to the phosphorus-loading concept in limnology. *Schweizerische Zeitschrift Hydrobiologie*, **37**, 53–84.

Vollenweider, R.A. (1976) Advances in defining critical loading levels for phosphorus in lake eutrophication. *Memorie dell'Istituto Italiano d'Idrobiologia*

Wang, J.K. (1990) Managing shrimp pond water to reduce discharge problems. *Aquacultural Engineering*, **9**, 61–73.

Watanabe, T. (1991) Past and present approaches to aquaculture waste management in Japan. In: *Nutritional Strategies and Aquaculture Waste* (eds C.B. Cowey & C.Y. Cho), pp. 137–54. University of Guelph, Ontario.

WCED (World Commission on Environment and Development) (1987) *Our Common Future*. Oxford University Press, Oxford, UK.

Weatherly, A. & Cogger, B. (1977) Fish culture: problems and prospects. *Science*, **197**, 427–40.

Westers, H. (1991) Operational waste management in aquaculture effluents. In: *Nutritional Strategies and Aquaculture Waste* (eds C.B. Cowey & C.Y. Cho), pp. 231–8. University of Guelph, Ontario.

Weston, D. (1986) *The Environmental Effects of Floating Mariculture in Puget Sound*. College of Ocean and Fishery Science, University of Washington, Seattle, WA.

Weston, D.P. (1991) The effects of aquaculture on indigenous biota. In: *Aquaculture and Water Quality*. Advances in World Aquaculture Vol. 3 (eds D.E. Brune & J.R. Tomasso), pp. 534–67. World Aquaculture Society, Baton Rouge.

Wilcox, D. (1994) Waste collection systems in net pen aquaculture. Presented at Aqua '94, Alexandria, MN, mimeo.

Wohlfarth, G. & Schroeder, G. (1979) Use of manure in fish farming – a review. *Agricultural Wastes*, **1**, 279–99.

Yarris, L. (1981) Trout manure as fertilizer. *Agriculture Research (USDA)*, **30**, 9.

Chapter 5
The Effects of Species Interactions Resulting from Aquaculture Operations

Angela H. Arthington & David R. Blühdorn   *Centre for Catchment and In-stream Research, Griffith University, Queensland, Australia.*

5.1   INTRODUCTION

With the declining returns of world fisheries, aquaculture is looked upon as a way to provide increasing quantities of aquatic product in the future (Beveridge 1987). Aquaculture production is increasing rapidly as modern techniques are applied to an expanding range of species (Pullin 1993a). Furthermore, subsistence and small-scale aquaculture is often the only source of animal protein for people in developing countries, as well as being a potential method for improving the standard of living of rural-based communities (Csavas 1993, Pullin 1993b). Thus aquaculture is important both as a means of providing food and for improving the quality of life.

Aquaculture activities and products from culture facilities can affect the environment in many ways. The principal adverse impacts include the destruction and fragmentation of natural habitats, changes in soil, water and landscape quality, changes in the abundance of species, impoverishment of genetic and biological diversity, and disturbance of ecosystem processes (GESAMP 1991, Weston 1991). The magnitude of such impacts varies with the nature and location of the culture system (extensive, semi-intensive, and intensive aquaculture in marine, brackish, and inland areas), the methods of husbandry used and the species under cultivation.

Irrespective of the type of aquaculture system or management strategies employed, escapes into the wild are virtually impossible to prevent (Beveridge & Phillips 1993). Therefore, all forms of aquaculture have one impact in common, the adverse ecological consequences to the indigenous biota and surrounding environment which may arise as the result of the escape of the organisms under culture (Pullin 1989; Chua 1993; Pullin 1993b).

There is now a general appreciation that species interactions, especially those resulting from the establishment of self-sustaining introduced species or the alteration of indigenous gene pools, are potentially the most damaging environmental consequences of aquaculture (Welcomme 1988; Barg 1992; Pullin *et al.* 1993). Whereas most of the effects of aquaculture on local habitats and water quality can be managed or minimised by careful selection of sites, effluent control and good husbandry, the

114

management of an established introduced species is extremely difficult. The effects of the ensuing species interactions may vary from regional to continental in scale and the impacts on indigenous biota are usually irreversible (Weston 1991; Pullin 1993b).

In this paper we review the major types of species interactions, with particular emphasis on the ecological relationships of exotic and indigenous species, and the consequences for aquatic ecosystems. Principles and examples are drawn from inland and coastal aquaculture systems, including intensive, semi-intensive, and extensive production in ponds, cages, pens, or other systems. Species interactions involving fish, molluscs and crustaceans are reviewed. The fundamental social and economic aspects driving aquaculture activity are also discussed as they relate to the various perspectives about species interactions and their impact on the environment.

The concept of aquaculture used in this paper is that defined by the FAO (1990), and comprises the farming of stock, by intervention in the rearing process to enhance production, under individual or corporate ownership. This paper is therefore not explicitly concerned with species which are introduced to create or enhance sport and harvest fisheries. Nevertheless, we stress that the environmental problems presented are relevant to the management of all introduced species irrespective of the motives for their introduction. In any case, the problems of exotic and translocated species in general are thoroughly documented elsewhere (e.g. Courtenay & Stauffer 1984; Bruton & van As 1986; Turner 1988; Welcomme 1988; De Silva 1989; Pollard 1990; Billington & Hebert 1991; Crowl *et al*. 1992).

An exotic species is defined as one that is not native to the country under discussion, while an indigenous species is one which is native to that country. The term 'introduced species' is used more generally to refer to any species intentionally or accidentally released into an environment outside its natural range (Welcomme 1988). Translocation refers to the movement of indigenous species to areas beyond their natural range but within the country of origin, and the movement of established exotic species to a new area. Welcomme (1988) used the terms 'transferred' and 'transplanted' to describe any species intentionally or accidentally transported and released within its previously described range, to enhance populations under stress or in decline, to introduce new genotypes or genetic diversity into a local stock, or to re-establish a species which has become locally extinct. An established species is one which has formed a self-sustaining population, and species described as invasive have demonstrated the ability to establish themselves in natural waters.

5.2 SPECIES INTRODUCTIONS FOR AQUACULTURE PURPOSES

In aquaculture, introductions of exotic species and translocations of indigenous species beyond their natural range are expressly intended to add entirely new elements (species or cultivars) to the production system. Aquaculture, especially intensive aquaculture, is a high risk enterprise and every advantage is taken of opportunities to enhance the productivity of stocks, the quality of the product, and profits. Species with a reputation for excellent performance under cultivation are the most likely choices for introduction to other areas, where building on past experience and available technologies can give a competitive edge and quick returns. Similarly, in the case of low-input aquaculture, species which offer the promise of a reliable source of protein, and which are cheaply available, are sought irrespective of their origins.

In addition, there have been many haphazard introductions of species for pilot aquaculture programs (Welcomme 1988), and a few species deliberately introduced as forage fish in aquaculture systems have escaped to the wild (e.g. species of gudgeon and minnow used in Spanish trout farms; Lobon-Cervia *et al.* 1989). In some cases, stocks deliberately introduced for aquaculture have been contaminated with other species which have subsequently escaped and established breeding populations. For example, the topmouth gudgeon, *Pseudorasbora parva* Schleegel, a south-east Asian cyprinid, was introduced accidentally in the Danube Delta in Romania in the 1960s. As well as achieving a pan-Danubian distribution (Rosecchi *et al.* 1993), this gudgeon has been transferred to Germany, Albania and Lithuania when stocking other species such as the European carp *Cyprinus carpio* L., and has also been reported in Israel, Italy, France and Greece (Rosecchi *et al.* 1993).

The global extent of species introductions associated with aquaculture has been reviewed comprehensively (Welcomme 1988; Baltz 1991; Munday *et al.* 1992a). In inland waters, introductions for aquaculture purposes appear to far exceed those for any other purpose, including introductions for sport fishing, improvement of wild stocks, trade in ornamental species, control of undesirable organisms (phytoplankton, plants, disease vectors and nuisance organisms such as mosquitoes) and accidental releases. Welcomme (1988) reported that 98 species of fish have been introduced internationally for aquaculture purposes involving inland waters.

Prior to 1900, the majority of fish species moved outside their normal ranges were salmonids, especially the rainbow trout *Oncorhynchus mykiss* (Walbaum), brown trout *Salmo trutta* L. and various species of *Salvelinus*. These species were introduced into temperate areas for cultivation and controlled releases to provide or enhance sport fisheries, and to a much lesser extent for food production. Salmonid introductions reached a peak in the 1890s, and more recently salmonid introductions have largely been limited to anadromous species which are being cultivated in mariculture systems (Welcomme 1988).

The movement of the European carp *Cyprinus carpio* began in Europe in medieval times, whilst more recent introductions peaked in the early decades of this century. Tilapia and Chinese carp have become predominant finfish culture species since the Second World War. The most recent wave of introductions has involved crustaceans, including shrimps and prawns for brackish water culture (Welcomme 1988) and crayfish in fresh water (Pillay 1992).

5.3   ESCAPES FROM AQUACULTURE

The potential for escape of exotic or translocated species from culture facilities has always been recognised as a risk in aquaculture developments. Escapes into the surrounding environment are inevitable in the long run, and they may involve very large numbers of individuals at any stage of the life history.

The subsequent fate of an introduced species in the new environment is generally unpredictable, since it will depend on dynamic interactions between the genetic, physiological and biological characteristics of the escapers and the characteristics, dynamics, and history of the receiving environment (Arthington & Mitchell 1986). Many of the characteristics inherent in aquaculture species, such as high reproductive success, wide environmental tolerances and broad habitat and dietary preferences, correspond to those which typify invasive species (Taylor *et al.* 1984; Bruton 1986).

Efforts to identify species likely to become self-sustaining and highly invasive, and to identify environments that are particularly susceptible to invasion (see Li & Moyle 1981; Bruton 1986; Welcomme 1988; Moyle *et al.* 1986; Crowl *et al.* 1992), have not really succeeded because of the strong element of chance and our limited understanding of the processes which regulate natural aquatic communities (Baltz 1991).

Another confounding factor is that many of the habitats invaded have been disturbed by human activities, and the significance of such disturbances is not well understood, although often invoked as favouring the establishment of exotic species, at least in freshwater systems (Arthington *et al.* 1990; Courtenay 1990; Crowl *et al.* 1992). Owing to the paucity of appropriate research, including before-and-after studies, many of the effects attributed to introduced species are not supported by conclusive data (Clugston 1990) and rely on conjecture to separate them from other causal factors such as habitat disturbance, nutrient enrichment and pollution. These factors frequently tend to be associated with aquaculture activities.

5.4   EFFECTS OF INTRODUCED SPECIES

Beveridge & Phillips (1993) summarise the potential adverse impacts of escaped aquaculture organisms into five categories:

- alterations to the host environment;
- disruption of the host community (principally through predation and competition);
- genetic degradation of local stocks;
- introduction of parasites and diseases;
- socio-economic effects.

Escapers do not have to reproduce in the new environment to cause an impact. The release of very large numbers of individuals which survive to feed and grow will have some effect on local resources and species. For example, typhoons in the Philippines regularly destroy fish pens, and one incident in 1976 released millions of milkfish *Chanos chanos* (Forsskal) into Laguna de Bay, boosting local harvest fisheries for weeks after the event (Gabriel 1979). Repeated escapes of exotic species which are regularly imported as larvae and juveniles for grow-out aquaculture are of particular concern, since large populations can persist without natural reproduction (Baltz 1991). A sustained predatory or competitive effect on indigenous species may follow the escape of long-lived species such as anguillids, which may survive for 30 years or more and not reproduce (Baltz 1991).

However, many exotic species escaping to the wild do reproduce and eventually become established and invasive in the new environment. According to Welcomme (1988), about two-thirds of the freshwater species introductions in the tropics have become successfully established. Species that remain a rare component of the aquatic community may have little impact, although this should not necessarily be assumed. If we reject the notion of vacant niches (see Herbold & Moyle 1986; Kikkawa & Anderson 1986), and recognise instead that the introduction of an additional species will result in the redistribution of resources amongst a portion of the community, then at least some change must occur. For rare species, the direction of these changes and their extent and time scales are for the most part obscure, and we have almost no knowledge of their functional effects.

When an established introduced species becomes predominant in the host community, more obvious and measurable changes may take place, ranging from effects on the local aquatic environment to severe disturbance of the community. We regard all of the effects of escaped aquaculture species, and the responses of indigenous biota to aquaculture facilities, as species interactions and review them individually below.

5.5   SPECIES INTERACTIONS

5.5.1   Disturbance of the local aquatic environment

There are surprisingly few good examples of environmental degradation due to escaped aquaculture organisms in the sense of direct effects on physical habitat, water quality and biological resources required by other biota. One of the most obvious nuisance species is the European carp *C. carpio*, which has been spread to at least 50 countries for cultivation as a food fish, as an ornamental species for ponds and lakes, and to enhance fisheries (Welcomme 1988). Wild populations have become established in many countries within the limits of the species' thermal tolerance, and in some areas introductions of carp have been beneficial. In others, including the United States, Europe, India, South Africa and Australia, the carp has acquired a reputation for causing the degradation of aquatic habitats and water quality (Crivelli 1983; Bruton & van As 1986; Moyle *et al.* 1986; Welcomme 1988; Fletcher *et al.* 1985).

Carp disturb the benthic sediments of freshwater lakes and slow-flowing rivers during feeding, disrupting the production of aquatic invertebrates (Moyle *et al.* 1986) and damaging aquatic macrophytes, especially delicate species (Crivelli 1983; Fletcher *et al.* 1985). The roiling behaviour of carp is believed to increase turbidity levels by re-suspending sediments (but see Fletcher *et al.* 1985), and the fish excrete nutrients which may contribute to accelerated eutrophication (Bruton 1985; Welcomme 1988) and possibly cyanobacterial outbreaks (P. Gherke, pers. comm.). In India, eutrophication and the shading out of macrophytes have led to changes in the composition of the indigenous fish fauna, including the disappearance of species in the genus *Schizothorax* together with their associated fisheries (Jhingran & Sehgal 1978).

Experiences with the European carp in Australia illustrate the full range of effects ascribed to this species elsewhere. Of the three varieties found in Australia, only one, the hybrid River (or Boolara) strain, an aquaculture escaper, has become invasive, and has undergone extensive range expansion in Australia's largest river system, the Murray-Darling. This has been accompanied, in many areas, by its domination of the fish community, contributing more than 80% of the total fish biomass in regions of the Murray, Lachlan and Murrumbidgee Rivers (P. Gehrke, pers. comm.). In addition to water quality impacts and a possible role in nutrient enrichment and the stimulation of cyanobacterial blooms, it is suspected that habitat modifications caused by carp have contributed to the decline of one endangered species, the trout cod *Maccullochella macquariensis* (Cuvier), and three vulnerable species, the dwarf galaxias *Galaxias pusilla* (Mack), the Yarra pygmy perch *Edelia obscura* (Klunzinger) and Ewen's pygmy perch *Nannoperca variegata* Kuiter and Allen (Water & Jackson 1993). The carp is thought to compete with several more common indigenous fishes for food (Fletcher *et al.* 1985). Finally, carp have been implicated as a secondary factor in the decline of native gastropods in the Murray River, South Australia. While river regulation is considered to have the greater impact, habitat alterations caused by carp may have changed the food available to indigenous aquatic snails (Sheldon & Walker 1993).

Grass carp *Ctenopharyngodon idella* (Valenciennes), although not introduced specifically for aquaculture purposes, is an herbivorous species that has had unforeseen adverse effects on the environment. By feeding selectively on more palatable species, it may shift the flora towards tougher species which are more of a nuisance than the plants originally targeted for control. There is also concern that the removal of plant beds may eliminate the spawning habit of phytophilous species, the refugia of young fish and amphibians, and the feeding habitat of certain water birds (Welcomme 1988).

Damage to physical habitat is less well established as an ecological impact on exotic species; structural damage has largely been reported because of its impact on human enterprises rather than natural resources. The Louisiana red crayfish *Procambarus clarkii* (Girard), indigenous to the United States, has been introduced for aquaculture purposes to Kenya, Uganda, the Sudan, Japan, parts of Europe, Hawaii and South and Central America (Welcomme 1988). In Europe it was introduced to replace *Astacus astacus* L. in natural waters after the devastation of this indigenous species by the European crayfish plague caused by the oomycete fungus *Aphanomyces astaci* L. *Procambarus* is regarded as a pest in many areas because of its burrowing behaviour, and the underground galleries it creates may cause extensive damage to earthen irrigation structures and the banks of aquaculture ponds. Its introduction to Japan as a food supply for bullfrogs resulted in damage to rice crops by its feeding on the plants and undermining of rice field dykes (Pillay 1992). The Chinese or mitten crab *Eriocheir sinensis* M. Edw. accidentally introduced into inland European waters in the ballast of ships, causes similar structural problems along river banks.

In Australia, the yabby *Cherax destructor* (Clark) has a wide natural distribution in central and southern inland areas. Recreational and commercial fisheries operate within its natural range and aquaculture activities are carried out both within the natural range and in Western Australia (Kailola *et al.* 1993). However, the Tasmanian Inland Fisheries Commission has declared the yabby a noxious species and opposes its introduction because of the crayfish's potential to damage irrigation channel and dam walls by burrowing, and to cause the deterioration of water quality in farm dams. There is also concern about the risk of disease transmission and competition with indigenous species of crayfish. On Kangaroo Island off the coast of South Australia, indigenous species of shrimps are rarely found where translocated yabby have become well established (P. Suter, pers. comm.).

There may be many more instances of aquatic habitat deterioration caused by exotic species, but such effects tend to be noticed mainly when they interfere with human property and production systems. It is also worth emphasising that the effects of introduced species on water quality and habitat are frequently masked by changes brought about by human activities. In the Murray-Darling River system, the direct impact of carp on turbidity levels has been difficult to distinguish from natural variations in turbidity associated with flooding and drying sequences, and from the effects of accelerated catchment and bank erosion on suspended solids levels (Fletcher *et al.* 1985).

5.5.2 Disturbance of the natural community

Disruption of the surrounding biotic community produced by species interactions associated with aquaculture operations may occur in a number of ways. Predation and competition are the principal processes involved. However, the effects of attraction to aquaculture facilities and the collection of wild seed or broodstock may also contribute significantly in some cases. Often the causal elements are unknown or obscured by other factors, such as pollution or habitat alteration. Causal processes rarely operate unilaterally, most being inseparably interrelated. Introduced piscivorous predators, for example, are reported to interact with indigenous biota by various types of competition (exploitation, interference, spatial) as well as predation, and these interactions will vary throughout the ontogeny of the predator.

The end result of such interactions, irrespective of the particular causal processes, is the impoverishment of diversity. For example, in the Philippines, the exotic catfish *Clarias batrachus* (L.) is reported to have displaced the indigenous catfish *C. macrocephalus* Günther (Juliano *et al.* 1989). In India, the indigenous carp *Catla catla* (Ham. Buch.) and *Labeo rohita* Hamilton are reported to have declined in certain reservoirs owing to the introduction of silver carp, *Hypophthalmichthys molitrix* (Valenciennes) (Shetty *et al.* 1989). In Malaysia, the snakeskin gouramy *Trichogaster pectoralis* (Regan) is reported to have displaced the indigenous congener, *T. trichopterus* (Pallas), to some extent (Ang *et al.* 1989).

5.5.3 Wild caught feed, broodstock and seed

Carnivorous species under culture conditions require large quantities of animal protein, and this is often supplied from wild caught stocks. For example, Iwama (1991) indicated that 6 kg of herring are needed to produce 1 kg of rainbow trout. Increasing demand for wild caught feed may lead to overfishing and conflicts with other users of the resource.

A number of aquaculture species are grown from wild-caught seed stock. The milkfish *Chanos chanos* and penaeid prawns are examples of these (Iwama 1991; Barg 1992; Phillips *et al.* 1993). In Bangladesh, the collection of wild carp fry for stocking freshwater fish ponds is reported to have contributed to the decline of fish stocks (Beveridge & Phillips 1993). Overexploitation of the wild caught resource is an ever-present possibility. At the same time, other, non-target species also suffer considerable losses which result from post-larvae harvesting. Studies have indicated that wasted bycatch (the fry and larvae of the non-target species) can be 10–50 times the biomass of the collected post-larvae (Macintosh & Phillips 1992).

Prawn culture also requires the collection of individuals in breeding condition for the production of nauplii. As these are animals of marketable size, their collection has led to competition between harvest fishery and aquaculture interests. Protective sanctions have been required in a number of countries, for example the Philippines and Indonesia, to protect stocks from overfishing (Lee & Wickins 1992).

5.5.4 Attraction to culture operations

Beveridge & Phillips (1993) indicated that aquaculture structures such as cages and pens may act as fish aggregation devices. In addition, many species are attracted to aquaculture operations by the excess food which is generally available in the vicinity (Weston 1991). Enriched conditions caused by excess food often produce a succession in the abundance and diversity of biota attracted to aquaculture operations. Such conditions can lead to enhanced populations of indigenous or escaped fishes in the areas surrounding aquaculture operations as much as 12 times higher than distant, unaffected sites (Iwama 1991; Weston 1991).

Predators attracted to aquaculture operations include birds, snakes, monitor lizards and turtles, fish dolphins, rodents, mustelids and bears (Beveridge 1984; Iwama 1991). Predators may be attracted to culture facilities by the shelter they provide, and by the increased abundance of food provided by the culture species themselves, by fouling organisms and by indigenous prey species. In their efforts to access caged stock, aggressive or large predators can cause structural damage to enclosures and so greatly increase the possibility of escapes (Iwama 1991; Munday *et al.* 1992a).

Disease outbreaks in cultivated stock may also be increased by predators attracted to aquaculture facilities. While bird attacks may often be unsuccessful, a not inconsiderable number of caged fish are wounded by such attacks. Under the normally crowded culture conditions, such damage increases the susceptibility of the fish to bacterial or fungal infections (Beveridge 1984; Iwama 1991). Predators may also act as intermediate hosts of parasites, or assist in the transfer of pathogens. In several cases in the United Kingdom, caged trout have developed severe infestations of the cestode *Diphyllobothrium*, resulting in heavy mortalities and the closure of one farm (Wootten 1979). The rapid spread of this parasite from its indigenous hosts was partly due to the migration of large numbers of gulls (*Larus* sp.) into the area (Beveridge 1984). Birds act as the intermediate host of the nematode *Contracaecum* sp., a common parasite of tilapia, as well as being responsible for many digenean infections of fish (Roberts & Sommerville, 1982).

5.5.5 Predation

Species interactions involving predation may be the most obvious (Courtenay 1990) and readily documented impact of exotic species, and they often result in the complete elimination of indigenous species in parts of their range. Globally, introduced salmonids and piscivorous species such as the largemouth bass *Micropterus salmonides* (Lacepede) are particularly notorious.

The rainbow trout *O. mykiss* is reported to be responsible for declines in indigenous fishes in Peru, Colombia, Chile, Yugoslavia, Himalayan rivers, South Africa and New Zealand (Welcomme 1988). In Lesotho, South Africa, *O. mykiss* preys on, and competes for food with, the rare indigenous minnow *Oreodaimon quathlambaep* (Barnard) (Bruton & van As 1986). The rainbow trout has been shown to prey on the Australian barred galaxis *Galaxias fuscus* Mack, an endangered species (Wager & Jackson 1993), and the distributions of *O. mykiss* and *Galaxias olidus* Günther in the Australian Capital Territory appear to be mutually exclusive (Lintermans 1991), presumably owing to predation. Trout have had similar impacts on the distribution of the common river galaxias, *G. vulgaris* Stokell in New Zealand streams (McDowall 1990). *O. mykiss* is also suspected of predation on at least two vulnerable indigenous fishes, *E. obscura* and *N. variegata* (Water & Jackson 1993). A paucity of indigenous fish species has been reported in areas where trout occur in the south-west areas of Western Australia (D. Morgan, pers. comm.).

The brown trout has also had a major impact on indigenous fish species and is implicated in the decline in numbers of four endangered and four vulnerable species and one species with a poorly known distribution in Australia (Wager and Jackson 1993). *S. trutta* is suspected of adversely interacting with the endangered Pedder galaxias *Galaxias pedderensis* Frankenberg, causing a dramatic decline in numbers. However, this decline is also linked to invasion by the translocated climbing galaxias *Galaxias brevipinnis* Günther (Wager & Jackson 1993).

Invasive predatory fishes, such as bass (*Micropterus* spp.) and trout (*O. mykiss* and *S. trutta*) have been implicated in the decline or local extinction of eight species of minnow (Cyprinidae), the Cape kurper *Sandelia capensis* (Cuvier) and *Kneria auriculata* (Pelligrin), some of which are classed as rare and endangered species in South Africa (Skelton 1993).

5.5.6   Competition

Exploitation competition is often invoked as the mechanism underlying the decline of indigenous fish species in areas where exotic species become established and abundant. This form of interaction occurs as a result of a shortage of some critical resource required by the competing organisms. The resource is usually food or space (i.e. the physical habitat required for spawning, foraging and other activities). Competition may alternatively involve a collection of effects termed interference, including territoriality, poisoning, injury or death by encounter (Schoener 1986) and inhibition of reproduction. The two types of competition are often imperfectly distinguished in descriptions of the interactions of exotic and indigenous species. Exploitation competition is notoriously difficult to demonstrate in the field, and most of the examples of impact attributed to competition have no experimental basis.

In spite of these difficulties, there is a strong belief that exotic species frequently out-compete indigenous species to the point of causing a considerable reduction in abundance, or even their complete disappearance. The brown trout is reported to have competed with, and displaced, indigenous salmonids in North America, and is actively excluded from some locations to facilitate the rehabilitation of populations of indigenous salmonids, including brook trout, *Salvelinus fontinalis* (Mitchill) and Atlantic salmon, *Salmo salar* L. (Clugston 1990).

The decline of indigenous fish species in Tashkent (former USSR) has been reported (Welcomme 1988, citing Rosenthal 1976) as a result of exotic species accidentally introduced with grass carp. Welcomme (1988, citing Noble 1980) noted that several indigenous species have been unable to compete with introduced tilapiine cichlids in southern USA. In the Tyume River, eastern Cape Province, South Africa, a rare species of kurper, *Andelia bainsi* Castelnau, is threatened by introduced rainbow trout, largemouth bass and the translocated sharptooth catfish *Clarias gariepinus* (Burchell), which compete for food and space (Bruton & van As 1986).

In parts of tropical Asia the tilapia *Oreochromis mossambicus* (Peters) contributes significant proportions of the animal protein available to local communities. In 1988 this species produced in excess of 100 000 t from capture fisheries and aquaculture operations (Petr 1992). However, in many Asian countries where this species has been introduced it is now considered a pest fish because of its invasive abilities, lack of social acceptance, and its propensity to overpopulate eutrophic waterbodies with masses of stunted individuals (Blühdorn & Arthington 1992). Stunted populations tend to crowd out established species in aquaculture systems and harvest fisheries by restricting living space, and in extreme cases may cause asphyxiation by creating an oxygen deficiency in the water column (Welcomme 1988). Several countries now regard *O. mossambicus* as unsuitable for culture (China and Malaysia) or as a pest (India, Taiwan and the Philippines) (De Silva 1989). Tilapia have interfered with the development of aquaculture in the Philippines and on several South Pacific islands (Nelson & Eldredge 1991).

Competition for breeding space has adversely affected the indigenous tilapia *O. variabilis* (Boulenger) in Lake Victoria, where introduced *Tilapia zillii* (Gervais) share the same nursery habitats (Welcomme 1988). Australian studies of the distribution and abundance of *O. mossambicus* have indicated the potential for competition with indigenous species for food and breeding territories, and stunting has occurred in both disturbed and relatively pristine habitats (Blühdorn *et al.* 1990; Arthington & Blühdorn 1994). *O. mossambicus* is considered to have the potential to devastate indigenous fish populations if it moves down the Darling River system from Queensland (P. Gehrke, pers. comm.), and is regarded as probably the most serious threat currently facing the Murray–Darling River system (B. Lawrence, pers. comm.)

In Australia, the decline and fragmentation of galaxiid populations has been attributed to interspecific competition with *S. trutta* for food (Fletcher 1979; Jackson & Williams 1980), and the blackfish *Gadopsis marmoratus* Richardson may have been similarly affected (Fletcher 1986), although in neither case was the role of predation entirely eliminated. Brown trout compete for food with the vulnerable indigenous Macquarie perch *Macquaria australasica* Cuvier & Valenciennes, and possibly prey on the juveniles of this species (Wager & Jackson 1993).

Mussels in large-scale farming systems in coastal lagoons, bays and inlets may compete with indigenous filter-feeders for planktonic food organisms and thus seriously affect their recruitment (Chua 1993). Suspended mussel culture in the Ria de Arosa, Spain, is reported to have replaced copepods as the principal pelagic grazing organism (Barg 1992). Intensive raft culture of the mussel *Mytilus edulis* L. in northwest Spain has changed the patterns of plankton composition and production, and the infaunal benthic community is affected by heavy organic enrichment from fecal wastes. However, the organic particulates move out onto the coastal shelf and support an enriched benthic community that may provide a significant food resource for demersal fishes (Tenore *et al.* 1985).

Interference competition has been described in relatively few instances, but the brown trout has been implicated in aggressive interactions with indigenous salmonids in the USA (Taylor *et al.* 1984).

5.5.7   Genetic interactions

Escaped aquaculture species may interact with indigenous species by breeding with local populations of the same species or through hybridisation with closely related species (Munday *et al.* 1992a; Beveridge & Phillips 1993). The escape of transgenic species from aquaculture facilities is regarded as a further dimension of the threat to indigenous biota arising from introduced species (Kapuscinski & Hallerman 1991).

Amongst the salmonids, there is good evidence of interbreeding between escapers from fish farms and local populations; for example, in southern Norwegian rivers, up to 28% of spawning Atlantic salmon *S. salar* may be of farmed origin (Munday *et al.* 1992a).

Wild Atlantic salmon populations show marked morphological differences between rivers, and local populations of salmonid fishes tend to be adapted to their specific environments (Munday *et al.* 1992a). Such adaptation is maintained by natal stream homing of the adult fish. Many traits in Atlantic salmon have a heritable genetic basis, including growth rate, age of maturation and smolting, egg size, timing of sea migration and migratory behaviour at sea (Institute of Aquaculture 1990). Thus there is concern that the adaptive traits and reproductive fitness of genetically distinct wild stocks may be significantly affected by interbreeding with introduced fish which escape from fish farms (Beveridge & Phillips 1993).

Studies in Sweden, France, Spain, Ireland, Canada and the USA have reported interbreeding of escaped brown trout and rainbow trout with indigenous populations; introgression rates of up to 80% have been recorded in France (Munday *et al.* 1992a). The observed effects of interbreeding vary from no measurable impact on the genetic structure of local stocks to partial or complete displacement of genetically distinct indigenous populations with homogeneous hatchery fish (Munday *et al.* 1992a).

Hybridisation may be a serious threat posed by both exotic and translocated aquaculture species, since interbreeding of closely-related species often produces viable offspring (Welcomme 1988). Hybridisation of Atlantic salmon and brown trout has been reported in Canada and Spain (Munday *et al.* 1992a) and in Australia under hatchery conditions (Fletcher 1986). Welcomme (1988) reported that the stresses associated with introduction may lead to a breakdown in normal behaviour and the formation of hybrids between species and even genera which do not normally hybridise when they coexist in the wild.

Interbreeding has occurred in Australia between two varieties of the European carp introduced for aquaculture, producing the vigorous Boolara strain which spread explosively in the 1960s and 1970s and became far more widespread and problematic than any of the original stocks (Shearer & Mulley 1978; Brumley 1991).

Much of the world's inland aquaculture uses carp and tilapiine cichlids and there has been widespread spontaneous and deliberate interbreeding of different genetic strains and species (Wohlfarth & Hulata 1983; Welcomme 1988). The wild genetic resources of both groups of fishes are believed to be threatened, and this may be true of catfish and other groups (Pullin 1993b). Localities identified in 1987 (Pullin 1988) for collection of pure stocks of tilapia in Africa have subsequently been found to contain fish of mixed origins, probably as a result of interbreeding with exotic stocks escaping from failed aquaculture ventures (Pullin 1993b). In a different instance, feral *O. mossambicus* were responsible for the commercial failure of the culture of all-male hybrids of *O. mossambicus* × *O. hornorum* (Trewavas) in Malaysia, because stray feral stock contaminated the hybrid stock, effectively eliminating any advantages of monosex culture (Ang *et al.* 1989).

In Taiwan, the exotic catfish, *C. batrachus* is reported to have hybridised with the indigenous congener *C. fuscus* Lacepede. This hybrid has spread over much of the island, reportedly to such an extent that the pure form of *C. fuscus* is in danger of extinction (Liao & Liu 1989). The conservation of wild genetic diversity in its own right and for future uses is steadily becoming a serious issue in many countries.

Beveridge & Phillips (1993) and Weston (1991) noted that there have been very few studies of genetic interactions between escaped aquaculture species and wild stocks. Assessment of the potential risk of adverse genetic effects has been attempted for salmonids, especially the Atlantic salmon (e.g. Hindar *et al.* 1991), although Munday *et al.* (1992a) considered that a better understanding of the genetics and population dynamics of wild Atlantic salmon is required before impacts due to interbreeding and loss of adaptiveness can be assessed.

Transgenic species have been produced for aquaculture, but little is known of their potential effects, and considerable research is required before they are widely used in aquaculture (Beveridge & Phillips 1993). The use of transgenic species in aquaculture and the potential transfer of genetic material to indigenous stocks and species introduces another element into the management of aquaculture species. However, Kapuscinski & Hallerman (1991) stated that the ecological impacts of both fertile and sterile transgenic fish will depend more on their overall phenotypic performance than on specific genetic constructs inserted into their genomes.

5.5.8 Introduction of diseases and parasites

The dissemination of disease agents and parasites has accompanied the introduction and translocation of fish, crustaceans and molluscs throughout the world, and the management of pathogens is a serious issue in all countries with a large investment in aquaculture. For example, European carp in North America are reported to harbour 170 parasites, of which 138 are exotic species; they include algae, fungi, protozoans, flatworms, tapeworms, leeches and crustaceans, most of which occur in crowded or aquaculture conditions (Clugston 1990). A summary of the pathogens which may be transferred with rainbow trout, Atlantic salmon, eels, oysters, mussels and lobsters is given by Munday *et al.* (1992a).

Certain pathogens are considered to affect only their original host species, genus or family, and so the introduction of an infected species within the group may threaten other members present in the receiving country, in aquaculture systems or in the wild. The bacterial causative agent of furunculosis was probably introduced into the United Kingdom from Denmark with brown trout, and spread through movements of farmed trout (Pillay 1992). It was subsequently imported to Norway via salmon smolts from Scotland and has spread to indigenous populations of salmonids (Egidius 1987). Wild Atlantic salmon populations in Norway have suffered massive mortalities and, in some areas, total eradication caused by the monogenean fluke *Gyrodactylus* sp., introduced from infected salmon hatcheries in Sweden (Munday *et al.* 1992a; Pillay 1992).

The re-introduction of the European flat oyster *Ostrea edulis* L. from North America spread the oyster parasite *Bonamia* sp., which devastated the European flat oyster industry (Barg 1992). The introduction of commercial prawn species was linked to the spread of pathogens such as infectious haematopoietic necrosis virus (IHNV) and Monodon bacillovirus (MBV) (Barg 1992). MBV was responsible for the collapse of prawn culture in Taiwan in 1988 (Kwei Lin 1989).

Increasingly there are incidences where taxon- or species-specific diseases are transmitted to unrelated hosts. Furunculosis was introduced to Victoria, Australia, in the 1970s via infected Japanese goldfish (Trust *et al.* 1980). This episode brought goldfish ulcer disease to cultured and wild Australian goldfish and carp populations, an issue of some significance for the aquarium industry and aquaculturists. It was followed by restrictions on the movements of goldfish within Australia as a protection against disease in important salmonid fisheries; Tasmania, for example, requires that imported goldfish be certified free of goldfish ulcer disease (Langdon 1990).

The spread of imported pathogens from their exotic hosts to indigenous species is of relevance to environmental protection and may exact a high ecological and economic cost. However, the evidence of impacts on indigenous species is limited (Munday *et al.* 1992a; Pillay 1992).

Langdon & Humphrey (1987) described a new viral disease of unknown origin, epizootic haematopoietic necrosis virus (EHNV), affecting cultured rainbow trout and feral redfin perch *Perca fluviatilis* L. in Australia. This disease is known to be highly pathogenic to several indigenous Australian fishes, including the silver perch *Bidyanus bidyanus* (Mitchell), mountain galaxias *G. olidus* and Macquarie perch *Macquaria australasica* Cuvier & Valenciennes, and to a lesser extent, Murray cod *Maccullochella peeli* (Mitchell) (Langdon 1989). The translocation of redfin perch and salmonids by angling and government bodies without health certification thus poses a threat to valuable indigenous fish stocks in the wild.

Massive mortalities of cultivated silver barramundi *Lates calcarifer* Block due to a picornia-like virus, BPLV (Glazebrook *et al.* 1990), have recently caused havoc to the industry in Queensland and the Northern Territory. BPLV has been diagnosed from barramundi in Australia (Glazebrook *et al.* 1990; Munday *et al.* 1992b), as well as from stocks in Thailand and Tahiti. Recent applications to establish growout facilities for silver barramundi within the Murray-Darling Basin have been refused because there is preliminary evidence that Macquarie perch, Murray cod and silver perch are susceptible to BPLV (J. Glazebrook, pers. comm.). Asymptomatic carriers of BPLV have been detected in barramundi from South Australian hatcheries and one of the urgent issues is the development and ready availability of a sensitive and specific test for the detection of the virus in asymptomatic fish.

Infectious agents may be more pathogenic to atypical hosts. Similarly, they may cause clinical disease only in atypical hosts. Such infectious agents become a problem when the typical host species come into contact with unusual hosts. Examples are whirling disease in rainbow trout, proliferative kidney disease of salmonids and the North American crayfish plague fungus. This fungus is only mildly pathogenic to the host crayfish but has devastated native European astacids.

Thompson (1990) illustrated the spread of the crayfish plague throughout Europe and discussed the pathology of the fungus and the ecological consequences of the loss of endemic crayfish species as a result of the plague. He cited the introduction of the plague vector animal, the North American signal crayfish *Pacifastacus leniusculus* Dana, as an outstanding example of deleterious translocations, and listed irreparable shifts in species diversity, ecosystem stress, and damaged traditional fisheries as major impacts of the fungus imported via this species. One of the side effects of the crayfish plague was that devastation of the indigenous Swedish crayfish *A. astacus* allowed macrophytes such as *Chara* spp. and *Elodea canadensis* Rich. to proliferate, resulting in the elimination of game fish habitat (Thompson 1990).

Parasites may also be transferred from exotic species to indigenous forms. Parasites introduced with the exotic Pacific oyster *Crassostrea gigas* L., including the Japanese oyster drill *Ocenebra japonica*, the oriental copepod *Mytilicola* sp. and the oyster flatworm *Pseudostylochus* sp., are reported to have had serious adverse impacts on oyster stocks on the west coast of the USA (Clugston 1990; Barg 1992). Moyle (1986) noted that indigenous Californian fishes seemed to be more heavily parasitised by exotic parasites (e.g. the anchor worm *Lernaea cyprinacea* L.) than exotic fishes. The anchor worm is now common in several native Australian fishes (Lloyd *et al.* 1987).

Disease organisms may be transmitted from wild populations to cultivated species (Roberts 1985) because the stressful nature of aquaculture may render cultivated species relatively susceptible to infections. Such interaction and those between exotic species and indigenous pathogens are not well understood (Munday *et al.* 1992a). The role of predators in the transmission of diseases and parasites from the wild to cultivated species is also poorly documented (e.g. see Beveridge 1984).

5.5.9 Socio-economic perspectives

The perception of species interaction effects has a firm foundation in socio-economic conditions. A farmer in a developed country may be looking to diversify production; the manager of a multi-million dollar sea-cage enterprise will look for high profitability; a subsistence fisherman in an undeveloped country may be simply trying to feed his family. Each of these people will have a different perspective on species interactions resulting from escapes from their culture enterprise or attractions to it. The first may view escapes from unfenced, littoral ponds as an 'act of God' and an occurrence for which there is little financial or regulatory incentive to guard against (Thompson 1990). The second may view the destruction of cages by storms as an engineering problem, to be solved by the sufficient input of resources. The third may view locally abundant exotic species as a gift, promising an improved food supply (Beveridge & Phillips 1993).

An example of contrasting perspectives on escaped aquaculture species is given by Clugston (1990). The carp *C. carpio* is generally considered a trash fish in the USA. However, carp persists under degraded habitat conditions in some urban areas and provide opportunities for recreational fishing which would otherwise not be available. The tilapia *O. mossambicus* is generally considered to be a pest throughout Asia (Welcomme 1988), although this is not the case in Sri Lanka (De Silva & Senaratne 1989). However, even in some of those countries which officially consider this species a pest, nuisance, or trash fish, feral stocks are an important source of food (Blühdorn & Arthington 1992). For example, in Indonesia, *O. mossambicus* is reported to be very useful for small-scale fish farmers and low income groups, but is considered a competitive trash fish in more intensive aquaculture (Eidman 1989). In the Asian region, *O. mossambicus* remained the dominant species in capture fisheries and provided a significant contribution to aquaculture production in 1988 (Petr 1992).

The Asia-Pacific region produced 84.5% of global aquaculture products in 1990 and, unlike post-industrial countries, most of this production was destined for local markets (Cvasas 1993). Such markets are, by nature, conservative and will often reject new or non-traditional products, thus generating considerable market resistance to otherwise nutritionally suitable products (Liao & Liu 1989; Cvasas 1993). Thus, as Welcomme (1988) explained, an unbiased, objective assessment of the effects of an introduced species may be impossible to achieve, and local perceptions of the costs and benefits of an introduction must be taken into account when considering such effects.

5.6 EFFECTS OF CULTURE SYSTEMS ON SPECIES INTERACTIONS

The likelihood of escape from aquaculture facilities is generally a function of the value of the product rather than of any regulatory prohibitions. Thus, intensive aquaculture operations tend to have fairly expensive and relatively effective means for preventing escape. On the other hand, subsistence aquaculture operations generally have few barriers to escape, since they often rely on locally-occurring organisms to provide seed stock. In the past, government agencies have actively supported the wide distribution of favoured species for semi-intensive and extensive culture irrespective of the organism's status as exotic.

5.6.1   Extensive culture

Extensive culture involves the rearing of organisms under relatively natural conditions of habitat and water quality. Generally, no supplementary feed is provided and stocking densities are low. Extensive culture is usually a supplement to other activities for income generation or subsistence, and is conducted principally in tropical areas using planktivorous species, because of the high natural productivity of these regions (Iwama 1991).

Extensive prawn culture consumes large areas of mangroves for low productivity returns (Phillips *et al.* 1993). It involves the passive recruitment of seed stock, which is often unpredictable and scarce as the resource succumbs to overuse and other pressures, such as pollution and destruction of mangroves (Macintosh & Phillips 1992).

In extensive culture there are few barriers to the exchange of species with the external environment. The implications for species interactions are twofold. Firstly, escaped organisms generally consist of the same stocks as those in the surrounding environment, since this is often their source, so the environmental impact of such escapes will be minimal, although disease transfer remains a potential problem. Secondly, because of this close interaction between the aquaculture environment and surrounding waters, any exotic which becomes feral will, sooner or later, find its way into the culture environment. The impact of such an event will depend on the perspective of the farmer. The invading feral tilapia *O. mossambicus* is reported to have disrupted the extensive brackish water farming of the milkfish *C. chanos* in the Philippines. However, *O. mossambicus* is now an established species in brackish water farms in that country (Juliano *et al.* 1989).

5.6.2   Semi-intensive culture

Semi-intensive culture involves the rearing of organisms under controlled habitat conditions, although artificial containers are rarely used (Iwama 1991). The diet is supplemented and stocking densities are elevated above natural levels. Semi-intensive systems are used to produce low-to-high value product, often in conjunction with other aquatic species (polyculture) or other animals and plants (integrated culture), mainly in tropical areas.

Semi-intensive culture is a favoured approach in developing countries as it treads the middle ground between the high capital costs and economic risks of intensive culture and the requirement for large areas of land for extensive culture (Pullin 1989; Phillips *et al.* 1993).

Species interactions arise in the collection of seed and broodstock from the wild, and the escape of exotic organisms into the local aquatic environment. The demand for seed stock in prawn aquaculture has given rise to commercial production of post-larvae, which may alleviate some of the demand on the wild stocks. However, Macintosh & Phillips (1992) reported that the poor quality of hatchery-reared post-larvae in some areas had resulted in the shunning of such stocks, with preference given to wild-caught post-larvae, again increasing the demand on a diminishing natural resource.

5.6.3 Intensive culture

Intensive culture involves the rearing of organisms under high stocking densities with active disease control measures. Diet is completely controlled, as are the habitat and water quality. Such systems aim to produce high-value product concomitant with the high investment costs and risks involved. High value means that there is a strong economic incentive not to lose stock. For example, the high value of the largemouth bass *M. salmoides* in Taiwan has prompted great care to be taken to prevent its escape into natural waters. These precautions have prevented the adverse environmental effects reported for this species elsewhere from occurring in Taiwan (Liao & Liu 1989).

However, when escapes from intensive operations do occur, the high stocking densities mean that many organisms are released into the outside environment. Similarly, the high stocking densities attract greater numbers of predators and the large amounts of unconsumed food attract scavengers.

Intensive culture is practised predominantly in the temperate waters of developed countries, where it is used to produce large volumes of high-value fin fish in cage, raceway and pond systems. In these locations the principal causes of stock losses are bad weather, floods, vandalism and marauding animals (mammals, piscivorous birds, mustelids). Species interactions arising from these operations including hybridisation of escapers with indigenous congeners, disturbance to the natural community through predation, competition and attraction to the aquaculture structures and excess food resources, and the spread of disease and parasites.

In tropical countries, the principal intensive culture operations are centred on prawn production. In developing countries, such enterprises often suffer disease and effluent problems under intensive culture conditions, and present significant economic barriers for small-scale farmers (Phillips *et al.* 1993).

5.7 MANAGEMENT OF SPECIES INTERACTIONS

In its broadest sense, the management of species interactions originating from aquaculture operations is the management of exotic or translocated species. There are a number of cogent reviews of this topic (Courtenay & Stauffer 1984; Bruton & van As 1986; Turner 1988; Welcomme 1988; De Silva 1989; Pollard 1990; Billington & Hebert 1991; Crowl *et al.* 1992). However, layers of political, social and economic policy intervene between the fundamental ecological impacts of invasive species and the management approaches actually applied to such problems.

In developed countries, the establishment and operation of aquaculture ventures are regulated, and regulations are enforced to a much greater extent than in developing countries (Csavas 1993). The management of species interactions in developing countries will therefore depend less on government-imposed sanctions and more on the availability of appropriate methodologies for aquaculture planning, guidelines for site selection, and public sector support for research, development and extension services (Csavas 1993; GESAMP 1991). The aim of aquaculture operations in all countries should be the development of sustainable systems that avoid environmental harm (Pullin 1993b).

5.7.1   Select appropriate scales of operation

While intensive culture is reasonably successful in developed countries, it is generally less so in developing countries where such highly capital-intensive operations are out of the reach of small-scale farmers who represent the most numerous and needy group of potential aquaculturists in this area of the world (Pullin 1989). Based on the premise that, 'Poverty and environmental conservation cannot co-exist', Pullin (1993a) recommended that small-scale (household/village) semi-intensive aquaculture systems are the most socially and environmentally desirable for developing countries.

The demands of expanding urbanisation in these countries will probably require the development of some large-scale operations, but most of the technical advice and policy formulation for aquaculture development should be attuned to the specific needs and opportunities of small-scale systems, rather than be constrained by foreign cultural biases (Pullin 1993a).

Two other factors which affect aquaculture in developing countries are the fallacies that indigenous species are inferior and not worth developing for local aquaculture, and that 'short-cutting' protocols on introductions is desirable because exotic species add prestige to aid-funded projects (Pullin 1993b).

5.7.2   Use sterile stock where possible

Organisms which have been chemically or chromosomally sterilised have been used to achieve specific purposes in aquaculture without most of the risks involved in using fertile specimens. Triploid grass carp, for example, have been used for vegetation control in North America (Clugston 1990). Much of modern tilapia culture is carried out using infertile stock.

5.7.3   Research the effects of escapers

Research into the effects of species interactions is scarce, especially in relation to aquaculture in developing countries, where it is, arguably, most needed (Pullin 1993a). There is also no generally applicable method for predicting the effects of escapes in any of the areas in which aquaculture is practised (Pullin 1993b). This lack is further exacerbated by the often inadequate time-frames of the research that is conducted (Pullin 1993b). For example, the European carp *C. carpio* was introduced to Australian waters some 100 years before its massive invasion of the Murray–Darling River system in Australia (Brumley 1991).

5.7.4   Improve quarantine measures

The introduction of the crayfish plague caused by the fungus *Aphanomyces astaci*, along with the exotic North American signal crayfish, to Britain was the result of poor quarantine mechanisms. Legislatively, there were no laws preventing the importation of live animals, and regulations governing quarantine and escape prevention at aquaculture sites were ineffective (Thompson 1990). Operationally, token measures to prevent escape of the crayfish were ineffective, as were quarantine measures to prevent the release of viable fungal spores from the aquaculture site (Thompson 1990). As well as devastating the European crayfish industry, this fungus is reported to be capable of infecting other crayfish species, such as the red claw *Cherax quadricarinatus* von Martens, an indigenous Australian species (Lee & Wickins 1992).

5.7.5   International protocols for introductions

The growing awareness of problems arising from the establishment in the wild of exotic species originally introduced for aquaculture has encouraged the investigation of indigenous species, especially in wealthier countries. Whilst this may prove of great benefit, both economically and in terms of environmental care, for the host country, the more promising species and successful culture systems are likely to be marketed internationally, and may be introduced eventually to other countries, setting in train new waves of movement. Here international codes of practice can be brought into play.

Protocols on introductions allow the potential risks and benefits of introductions to be compared, and decisions to be made in the light of existing scientific knowledge (Coates 1993). The adoption of precautionary policies and codes of practice, such as those proposed by the FAO (Anon. 1994) and others (Neal 1984; Turner 1988), and their widespread implementation, will be a major step in developing a standardised approach to introductions and in facilitating a measure of consensus about likely species interactions. However, this will mean that hard decisions will have to be made about the traffic in promising new candidates for aquaculture, at the cost of lost profits for the country of origin.

5.8 SUMMARY

In conclusion, it is evident that a number of factors concerning species interactions arising from aquaculture activities are common across all types of operations. The two most fundamental elements can be the most simply stated: escapes are inevitable and invasions are irreversible. Thus, any cultured organism is a potentially invasive species.

The impacts of species interactions associated with aquaculture have, in some cases, resulted in alterations to the host environment and disruptions to the host community, which result in impoverishment of diversity, genetic disturbances, or the introduction of parasites and diseases. The effects of these impacts are filtered through socio-economic factors to produce often divergent perspectives about escaped organisms in particular, and aquaculture in general. These local perspectives are vital indicators of the types of management, scale of operations, and effectiveness of the regulatory sanctions which can be applied successfully.

Internationally, protocols exist to help determine whether a species should be introduced to a new environment. These should be adhered to rigorously and improved as time and experience determine. Nationally and locally, the risks involved in escapes from (or attractions to) aquaculture operations can be minimised by careful site selection, appropriate containment facilities and operations geared for sustainability, with environmentally responsible approaches applied to the stocked organisms and their husbandry, quarantine, effluent control and disease management.

REFERENCES

Ang, K.J., Gopinath, R. & Chua, T.E. (1989) The status of introduced fish species in Malaysia. In: *Exotic Aquatic Organisms in Asia* (ed. S.S. De Silva), pp. 71–82. *Asian Fisheries Society Special Publications*, **3**. Asian Fisheries Society, Manila, Philippines.

Anon. (1994) Guidelines for Responsible Aquaculture Development. Technical Consultation on the Code of Conduct for Responsible Fishing. FI: CCRF/94/Inf., **7**, FAO, Rome (first draft).

Arthington, A.H. & Blühdorn, D.R. (1994) Distribution, genetics, ecology and status of the introduced cichlid, *Oreochromis mossambicus*, in Australia. *Mitteilungen (Communications), Societas Internationalis Limnologiae (WIL)*, **24**, 53–62.

Arthington, A.H. & Mitchell, D.S. (1986). Aquatic invading species. In: *Ecology of Biological Invasions. An Australian Perspective* (eds R.H. Grooves & J.J. Burdon), pp. 34–53. Australian Academy of Science, Canberra.

Arthington, A.H., Hamlet, S. & Blühdorn, D.R. (1990) The role of habitat disturbance in the establishment of introduced warm-water fishes in Australia. In: *Introduced and Translocated Fishes and their Ecological Effects* (ed. D.A. Pollard), pp. 61–6. *Bureau of Rural Resources Proceedings*, **8**. Australian Government Publishing Service, Canberra.

Baltz, D.M. (1991) Introduced fishes in marine systems and inland seas. *Biological Conservation*, **56**, 151–77.

Barg, U.C. (1992) Guidelines for the promotion of environmental management of coastal aquaculture development. *FAO Fisheries Technical Paper*, **328**. FAO, Rome.

Beveridge, M.C.M. (1984) Cage and pen fish farming. Carrying capacity models and environmental impact. *FAO Fisheries Technical Paper*, **255**. FAO, Rome.

Beveridge, M.C.M. (1987) *Cage Aquaculture*. Fishing News Books, Farnham.

Beveridge, M.C.M. & Phillips, M.J. (1993) Environmental impact of tropical inland aquaculture. In: *Environment and Aquaculture in Developing Countries* (eds R.S.V. Pullin, H. Rosenthal & J.L. Maclean), pp. 213–36. *ICLARM Conference Proceedings*, **31**. International Center for Living Aquatic Resource Management, Manila, Philippines.

Billington, N. & Hebert, P.D.N. (eds) (1991) The ecological and genetic implications of fish introductions (FIN). *Canadian Journal of Fisheries and Aquatic Sciences*, **48** (Supplement 1), 80–94.

Blühdorn, D.R. & Arthington, A.H. (1992) Tilapia in Australia and small Pacific islands: unwanted pest or unappreciated resource? *Papers contributed to the Workshop on Tilapia in Capture and Culture-based Fisheries and Country Reports presented at the Fifth Session of the Indo-Pacific Fishery Commission Working Party of Experts on Inland Fisheries*. Bogor, Indonesia 24–9 June 1991 (ed E.A. Balayut), pp. 120–38. *FAO Fisheries Report*, **458** (Supplement). FAO, Rome.

Blühdorn, D.R., Arthington, A.H. & Mather, P.B. (1990) The introduced cichlid, *Oreochromis mossambicus*, in Australia: a review of distribution, population genetics, ecology, management issues and research priorities. In: *Introduced and Translocated Fishes and their Ecological Effects* (ed. D.A. Pollard), pp. 83–92. *Bureau of Rural Resources Proceedings*, **8**. Australian Government Publishing Service, Canberra.

Brumley, A.R. (1991) Cyprinids of Australasia. In: *Cyprinid Fishes – Systematics, Biology and Exploitation* (eds I.J. Winfield & J.S. Nelson), pp. 264–83. Chapman and Hall, London.

Bruton, M.N. (1985) Effects of suspensoids on fish. In: *Perspectives in Southern Hemisphere Limnology. Developments in Hydrobiology*, **28** (eds B.R. Davies & R.D. Walmsley), pp. 221–41. Dr. W. Junk Publishers, Dordrecht.

Bruton, M.N. (1986) Life history styles of invasive fishes in southern Africa. In: *The Ecology and Management of Biological Invasions in Southern Africa* (eds I.A.W. Macdonald, F.J. Kruger & A.A. Ferrar), pp. 201–8. Oxford University Press, Cape Town.

Bruton, M.N. & van As, J.G. (1986) Faunal invasions of aquatic ecosystems in southern Africa, with suggestions for their management. In: *The Ecology and Management of Biological Invasions in Southern Africa* (eds I.A.W. Macdonald, F.J. Kruger & A.A. Ferrar), pp. 47–61. Oxford University Press, Cape Town.

Chua, T.E. (1993) Environmental management of coastal aquaculture development. In: *Environment and Aquaculture in Developing Countries* (eds R.S.V. Pullin, H. Rosenthal & J.L. Maclean), pp. 199–212. *ICLARM Conference Proceedings*, **31**. International Center for Living Aquatic Resource Management, Manila, Philippines.

Clugston, J.P. (1990) Exotic animals and plants in aquaculture. *Reviews in Aquatic Science*, **2**, 481–9.

Coates, D. (1993) Environmental management implications of aquatic species introductions: a case study of fish introductions into the Sepik-Ramu Basin, Papua, New Guinea. *Asian Journal of Environmental Management*, **1**, 39–49.

Courtenay, W.R. Jr (1990). Fish introductions and translocations, and their impacts in Australia. In: *Introduced and Translocated Fishes and their Ecological Effects* (ed. D.A. Pollard), pp. 171–9. *Bureau of Rural Resources Proceedings*, **8**. Australian Government Publishing Service, Canberra.

Courtenay, W.R. Jr. & Stauffer, J.R., Jr (eds) (1984). *Distribution, Biology and Management of Exotic Fishes*. Johns Hopkins University Press, Baltimore, Maryland.

Crivelli, A.J. (1983) The destruction of aquatic vegetation by carp. *Hydrobiologia*, **106**, 37–41.

Crowl, T.A., Townsend, C.R. & McIntosh, A.R. (1992). The impact of introduced brown and rainbow trout on native fish: the case of Australasia. *Reviews in Fish Biology and Fisheries*, **2**, 217–41.

Csavas, I. (1993) Aquaculture development and environmental issues in the developing countries of Asia. In: *Environment and Aquaculture in Developing Countries* (eds R.S.V. Pullin, H. Rosenthal & J.L. Maclean), pp. 74–101. *ICLARM Conference Proceedings*, **31**. International Center for Living Aquatic Resource Management, Manila, Philippines.

De Silva, S.S. (ed.) (1989) *Exotic Aquatic Organisms in Asia. Asian Fisheries Society Special Publication*, **3**, Asian Fisheries Society, Manila, Philippines.

De Silva, S.S. & Senaratne, K.A.D.W. (1989) *Oreochromis mossambicus* is not universally a nuisance species: the Sri Lankan experience. In: *The Second International Symposium on Tilapia in Aquaculture* (eds R.S.V. Pullin, T. Bhukaswan, K. Tonguthai & J.L. Maclean), pp. 445–50. *ICLARM Conference Proceedings*, **15**. Department of Fisheries, Bangkok, Thailand, and International Center for Living Aquatic Resources Management, Manila, Philippines.

Egidius, E. (1987) Import of furunculosis to Norway with Atlantic salmon smolts from Scotland. ICES CM 1987/F: **8**.

Eidman, H.M. (1989) Exotic aquatic species introduction into Indonesia. In: *Exotic Aquatic Organisms in Asia* (ed. S.S. De Silva), pp. 57–62. *Asian Fisheries Society Special Publication*, **3**. Asian Fisheries Society, Manila, Philippines.

FAO (1990) The definition of aquaculture and collection of statistics. *Aquaculture Minutes*, **7**. Inland Water Resources and Aquaculture Services. Fisheries Department, FAO, Rome.

Fletcher, A.R. (1979) *Effects of* Salmo trutta *on* Galaxias olidus *and macroinvertebrates in stream communities*. M.Sc. thesis, Monash University, Victoria.

Fletcher, A.R. (1986) Effects of introduced fish in Australia. In: *Limnology in Australia* (eds P. De Deckker & W.D. Williams), pp. 231–8. CSIRO, Melbourne and Dr W. Junk, Dordrecht.

Fletcher, A.R., Morison, A.K. & Hume, D.J. (1985) Effects of carp, *Cyprinus carpio* L, on communities of aquatic vegetation and turbidity of waterbodies in the lower Goulburn River basin. *Australian Journal of Marine and Freshwater Research*, **36**, 311–27.

Gabriel, B.C. (1979) Milkfish culture in freshwater pens. In: *Technical Consultation on Available Aquaculture Technology in the Philippines*, pp. 114–19. Aquaculture Department SEAFDEC & Philippine Council for Agriculture and Resources Research. AQB/SEAFDEC, Iloilo, Philippines.

GESAMP (1991) *Reducing Environmental Impacts of Coastal Aquaculture*. Reports and Studies GESAMP (IMO/FAO/Unesco/WMO/WHO/IAEA/UN/UNEP Joint Group of Experts on the Scientific Aspects of Marine Pollution), 47. FAO, Rome.

Glazebrook, J.S., Heasman, M.P. & de Beer, S.W. (1990). Picornia-like viral particles associated with mass mortalities in larval barramundi, *Lates calcarifer* Bloch. *Journal of Fish Diseases*, **13**, 245–9.

Herbold, B. & Moyle, P.B. (1986) Introduced species and vacant niches. *The American Naturalist*, **128**, 751–60.

Hindar, K., Ryman, N. & Utter, F. (1991) Genetic effects of aquaculture on natural fish populations. *Aquaculture*, **98**, 259–62.

Institute of Aquaculture (1990) *Fish Farming and the Scottish Freshwater Environment*. Report prepared for the Nature Conservancy Council by: Institute of Aquaculture, University of Stirling; Institute of Freshwater Ecology, Bush Estate, Penicuik; and Institute of Terrestrial Ecology, Banchory. Nature Conservancy Council, Edinburgh.

Iwama, G.K. (1991) Interactions between aquaculture and environment. *Critical Reviews in Environmental Control*, **21**, 177–216.

Jackson, P.D. & Williams, W.D. (1980) Effects of brown trout, *Salmo trutta* L., on the distribution of some native fishes in three areas of southern Victoria. *Australian Journal of Marine and Freshwater Research*, **31**, 61–7.

Jhingran, V.G. & Sehgal, K.L. (1978) *The Cold Water Fisheries of India*. Barrackpore, West Bengal. Inland Fisheries Society of India.

Juliano, R.O., Guerrero, R. III & Ronquillo, I. (1989) The introduction of exotic aquatic species in the Philippines. In: *Exotic Aquatic Organisms in Asia* (ed. S.S. De Silva), pp. 83–90. *Asian Fisheries Society Special Publication*, **3**. Asian Fisheries Society, Manila, Philippines.

Kailola, P.J., Williams, M.J., Stewart, P.C., Reichelt, R.E., McNee, A. & Grieve, C. (1993) *Australian Fisheries Resources*. Bureau of Resource Sciences and Fisheries Research and Development Corporation, Canberra.

Kapuscinski, A.R. & Hallerman, E.M. (1991) Implications of introduction of transgenic fish into natural ecosystems. *Canadian Journal of Fisheries and Aquatic Science*, **48** (Supplement 1), 110–17.

Kikkawa, J. & Anderson, D.J. (eds) (1986) *Community Ecology: Pattern and Process*. Blackwell Scientific Publications, Melbourne.

Kwei Lin, C. (1989) Prawn culture in Taiwan: What went wrong? *World Aquaculture*, **20**, 19–20.

Langdon, J. (1989) Experimental transmission and pathogenicity of epizootioc haematopoietic necrosis virus (ENHV) in redfin perch, *Perca fluviatilis* L., and 11 other teleosts. *Journal of Fish Diseases*, **12**, 295–310.

Langdon, J.S. (1990) Disease risks of fish introductions and translocations. In: *Introduced and Translocated Fishes and their Ecological Effects* (ed. D.A. Pollard), pp. 98–107. *Bureau of Rural Resources Proceedings*, **8**. Australian Government Publishing Service, Canberra.

Langdon, J. & Humphrey, J.D. (1987) Epizootic haematopoietic necrosis, a new viral disease in redfin perch, *Perca fluviatilis* L., in Australia. *Journal of Fish Diseases*, **10**, 289–97.

Lee, D. O'C. & Wickins, J.F. (1992) *Crustacean Farming*. Blackwell Scientific Publications, Oxford.

Li, H.W. & Moyle, P.B. (1981) Ecological analysis of species introductions into aquatic systems. *Transactions of the American Fisheries Society*, **110**, 772–82.

Liao, I.C. & Liu, H.C. (1989) Exotic aquatic species in Taiwan. In: *Exotic Aquatic Organisms in Asia* (ed. S.S. De Silva), pp. 101–18. Asian Fisheries Society Special Publication, **3**. Asian Fisheries Society, Manila, Philippines.

Lintermans, M. (1991) The decline of native fish in the Canberra Region: the impacts of introduced species. *Bogong*, **12**, 18–22.

Lloyd, L. Arthington, A.H. & Milton, D.A. (1987) The mosquitofish, *Gambusia affinis*, a valuable mosquito control agent or a pest? In: *The Ecology of Exotic Animals and Plants in Australasia* (ed. R.L. Kitching), pp. 6–25. Jacaranda Wiley Press, Brisbane.

Lobon-Cervia, J., Elvira, B. & Rincon, P.A. (1989) Historical changes in the fish fauna of the River Duero Basin. In: *Historical Change of Large Alluvial Rivers: Western Europe* (eds G.A. Petts, H. Moller & A.L. Roux), pp. 221–32. John Wiley and Sons, Chichester.

Macintosh, D.J. & Phillips, M.J. (1992) Environmental issues in shrimp farming. In: *Shrimp '92* (eds H. De Saram & T. Singh), pp. 118–45. *Proceedings of the 3rd Global Conference on the Shrimp Industry*, Hong Kong, 14–16 September 1992. INFOFISH, Malaysia.

McDowall, R.M. (1990) When galaxiid and salmonid fishes meet – a family reunion in New Zealand. *Journal of Fish Biology*, **37** (Supplement A), 35–43.

Moyle, P.B. (1986) Fish introductions into North America: patterns and ecological impact. In: *Ecology of Biological Invasions in North America and Hawaii* (eds H.A. Mooney & J.A. Drake), pp. 27–43. Springer-Verlag, New York.

Moyle, P.B., Li, H.W. & Barton, B.A. (1986) The Frankenstein effect: impact of introduced fishes on native fishes in North America. In: *Fish Culture in Fisheries Management* (ed. R.H. Stroud), pp. 415–26. American Fisheries Society, Bethesda.

Munday, B., Eleftheriou, A. Kentouri, M. & Divanach, P. (1992a). *The Interactions of Aquaculture and the Environment – A Bibliographical Review*. Commission of the European Communities, Directorate-General of Fisheries.

Munday, B.L., Langdon, J.S., Hyatt, A. & Humphrey, J.D. (1992b) Mass mortality associated with a viral-induced vacuolating encephalopathy and retinopathy of larval and juvenile barramundi, *Lates calcarifer* Bloch. *Aquaculture*, **103**, 197–211.

Neal, R.A. (1984) Aquaculture expansion and environmental considerations. *Mazingira*, **8**, 24–8.

Nelson, S.G. & Eldredge, L.G. (1991) Distribution and status of introduced cichlid fishes of the genera *Oreochromis* and *Tilapia* in the Islands of the South Pacific and Micronesia. *Asian Fisheries Science*, **4**, 11–22.

Noble, L.E. (1980) *The history, status and identification of cichlid fishes in the southern United States*. Dissertation, College Station, Texas A and M University.

Petr, T. (1992) Tilapia in capture and culture-enhanced fisheries of the Indo-Pacific. In: *Papers Contributed to the Workshop on Tilapia in Capture and Culture-Based Fisheries and Country Reports presented at the Fifth Session of the Indo-Pacific Fishery Commission Working Party of Experts on Inland Fisheries*. Bogor, Indonesia 24–29 June 1991 (ed. E.A. Balayut), pp. 115–19. *FAO Fisheries Report*, **458** (Supplement). FAO, Rome.

Phillips, M.J., Kwei Lin, C. & Beveridge, M.C.M. (1993) Shrimp culture and the environment: lesions from the world's most rapidly expanding warmwater aquaculture sector. In: *Environment and Aquaculture in Developing Countries* (eds R.S.V. Pullin, H. Rosenthal & J.L. Maclean), pp. 171–97. *ICLARM Conference Proceedings*, **31**. International Center for Living Aquatic Resource Management, Manila, Philippines.

Pillay, T.V.R. (1992) *Aquaculture and the Environment*. Fishing News Books, Oxford.

Pollard, D.A. (ed.) (1990) *Introduced and Translocated Fishes and their Ecological Effects. Bureau of Rural*

Resources Proceedings, **8**. Australian Government Publishing Service, Canberra.

Pullin, R.S.V. (ed.) (1988) *Tilapia Genetic Resources for Aquaculture. ICLARM Conference Proceedings*, **16**. International Center for Living Aquatic Resource Management, Manila, Philippines.

Pullin, R.S.V. (1989) Third-world aquaculture and the environment. *Naga*, **12**, 10–13.

Pullin, R.S.V. (1993a) An overview of environmental issues in developing-country aquaculture. In: *Environment and Aquaculture in Developing Countries* (eds R.S.V. Pullin, H. Rosenthal & J.L. Maclean), pp. 1–19. *ICLARM Conference Proceedings*, **31**. International Center for Living Aquatic Resource Management, Manila, Philippines.

Pullin, R.S.V. (compiler) (1993b) Discussion and Recommendations on Aquaculture and the Environment in Developing Countries. In: *Environment and Aquaculture in Developing Countries* (eds R.S.V. Pullin, H. Rosenthal & J.L. Maclean), pp. 312–38. *ICLARM Conference Proceedings*, **31**. International Center for Living Aquatic Resource Management, Manila, Philippines.

Pullin, R.S.V., Rosenthal, H. & Maclean, J.L. (eds) (1993) *Environment and Aquaculture in Developing Countries. ICLARM Conference Proceedings*, **31**. International Center for Living Aquatic Resource Management, Manila, Philippines.

Roberts, M.S. (1985) Why ERM gives cause for concern. *Fish Farmer*, **8**, 27.

Roberts, R.J. & Sommerville, C. (1982) Diseases of tilapia. In: *The Biology and Culture of Tilapia* (eds R.S.V. Pullin & R.H. Lowe-McConnell), pp. 247–63. *ICLARM Conference Proceedings*, **7**. International Center for Living Aquatic Resources Management, Manila, Philippines.

Rosecchi, E., Crivelli, A.J. & Catsadorakis, G. (1993) The establishment and impact of *Pseudorasbora parva*, an exotic fish species introduced into Lake Mikri Prespa (north-western Greece). *Aquatic Conservation: Marine and Freshwater Ecosystems*, **3**, 223–31.

Rosenthal, H. (1976) Implications of transplantations to aquaculture and ecosystems. *Paper presented to the FAO Conference on Aquaculture, Kyoto, Japan, 26 May–2 June 1975*. FAO, Rome.

Schoener, T.W. (1986) Resource partitioning. In: *Community Ecology: Pattern and Process* (eds J. Kikkawa & D.J. Anderson), pp. 91–126. Blackwell Scientific Publications, Melbourne.

Shearer, K.D. & Mulley, J.C. (1978) The introduction and distribution of the carp, *Cyprinus carpio* Linnaeus, in Australia. *Australian Journal of Marine and Freshwater Research*, **29**, 551–64.

Sheldon, F. & Walker, K.F. (1993) Pipelines as a refuge for freshwater snails. *Regulated Rivers: Research and Management*, **8**, 295–9.

Shetty, H.P.C., Nandeesha, M.C. & Jhingran, A.G. (1989) Impact of exotic aquatic species in Indian waters. In: *Exotic Aquatic Organisms in Asia* (ed. S.S. De Silva), pp. 45–55. *Asian Fisheries Society Special Publication*, **3**. Asian Fisheries Society, Manila, Philippines.

Skelton, P.H. (1993) *A Complete Guide to the Freshwater Fishes of Southern Africa*. Southern Book Publishers, Halfway House, South Africa.

Taylor, J.N., Courtenay, W.R., Jr. & McCann, J.A. (1984) Known impacts of exotic fishes in the continental United States. In: *Distribution, Biology and Management of Exotic Fishes* (eds W.R. Courtenay, Jr. & J.R. Stauffer), pp. 322–73. Johns Hopkins University Press, Baltimore.

Tenore, K.R., Coral, J., Gonzalez, N. & Lopez-Jamar, E. (1985) Effects of intense mussel culture on food chain patterns and production in coastal Galicia, NW Spain. In: *Proceedings of the International Symposium on Utilisation of Coastal Ecosystems: Planning, Pollution, and Productivity*, Vol. 1 (eds N. Labish-Chao & W. Kirby-Smith), pp. 321–8.

Thompson, A.G. (1990) The danger of exotic species. *World Aquaculture*, **21**, 25–32.

Trust, T.J., Khouri, A.G., Austin, R.A. & Ashburner, L.D. (1980) First isolation in Australia of atypical *Aeromonas salmonicida*. *FEMS Microbiology Letters*, **9**, 39–42.

Turner, G.E. (ed.) (1988) *Codes of Practice and manual of Procedures for Consideration of Introductions and Transfers of Marine and Freshwater Organisms. EIFAC Occasional Paper*, **23**. FAO, Rome.

Wager, R. & Jackson, P. (1993) *The Action Plan for Australian Freshwater Fishes*. Australian Nature Conservation Agency, Canberra.

Welcomme, R.L. (1988). *International Introductions of Inland Aquatic Species. FAO Fisheries Technical Paper*, **294**. FAO, Rome.

Weston, D.P. (1991) The effects of aquaculture on indigenous biota. In: *Aquaculture and Water Quality. Advances in World Aquaculture*, Vol. 3 (eds D.E. Brune & J.R. Tomasso), pp. 535–67. World Aquaculture Society, Baton Rouge, Louisiana, USA.

Wohlfarth, G.W. & Hulata, G. (1983) *Applied Genetics of Tilapia*, 2nd edn. *ICLARM Studies and Reviews*, **6**. International Center for Living Aquatic Resources Management, Manila, Philippines.
Wooten, R. (1979) Tapeworm threat to trout in floating freshwater cages. *Fish Farmer*, **2**, 5.

Chapter 6
Environmental Considerations in the Use of Antibacterial Drugs in Aquaculture

Donald P. Weston   *Department of Integrative Biology, University of California, Berkeley, California, USA*

6.1 INTRODUCTION

The use of antibacterial agents in terrestrial agriculture is accepted as standard industry practice. Antibacterials and parasiticides are routinely given as feed additives to virtually all species of domesticated animals both to prevent and to control disease. In the United States alone, over 250 million US dollars is spent annually for veterinary use of antibacterial feed supplements (Stinson 1987). Thus it is not surprising that aquaculturists find a need for antibacterial chemotherapy as well, but this use has provoked concern in scientific circles (Michel 1986; Brown 1989) and among the general public (Stickney 1988; Barinaga 1990).

Although there is some concern regarding environmental consequences of antibiotic use in all forms of animal production, this concern appears to be particularly acute in aquaculture. Three factors may contribute to this heightened awareness.

First, widespread aquacultural use of antibacterials is a relatively new practice. Although sulphanilimide was used (unsuccessfully) to treat a *Hemophilus piscium* infection in brook trout in 1937 (Tunison & McCay 1937), the use of aquacultural antibacterials was negligible until the 1980s. This decade was characterised by rapid increases in production in many industry segments, including salmonids in Norway and elsewhere, penaeid shrimp in Asia and parts of South America, and catfish in the United States. All these industry segments rely upon antibacterials to some degree, and it is likely that drug usage has increased concomitantly with biomass production.

Secondly, in aquaculture systems there is often little or no effluent treatment and the water provides immediate transport of antibacterial residues between the farm and the surrounding environment. In net-cage culture, for example, the farm boundary is defined by nothing more than an open-meshed net. Therefore, any drug not retained or metabolised by the cultured organism can be directly and immediately released to the environment beyond the culture facility.

Finally, the paucity of reliable scientific data on the aquatic fate and effects of antibacterials has contributed to the current state of uncertainty. In many developing

140

countries aquacultural drugs are used with little regulatory control or consideration of environmental effects. Even in countries with strict licensing procedures for aquacultural chemotherapeutants, the emphasis historically has been on efficacy of the drug and safety to the human consumer. Environmental data requirements to obtain product approval have been minimal, and thus neither the manufacturers nor users have had any impetus to acquire information on therapeutant susceptibility to microbial degradation or photolysis, propensity to bioaccumulate, stimulation of resistance in non-target microbes, etc. While this situation is changing for drugs now entering the approval process, we are currently in the position of having many drugs in widespread use for which we know little or nothing about potential environmental fate and effects.

It is the purpose of this paper to synthesise available information on the environmental fate and effects of antibacterial drugs in aquaculture in order to identify and assess the potential risks. In addition, data are presented from a recently completed study of the environmental effects of antibacterial use in salmonid net-cage culture in the north-western United States (Weston *et al.* 1994). The environmental issues considered are: (1) persistence of drug residues in water and sediments, (2) appearance of drug residues in non-target organisms, (3) stimulation of antibacterial resistance, (4) effects on microbially-mediated biogeochemical process, and (5) human health issues.

In order to discuss environmental issues in proper context, information will first be presented on standard industry practices for antibacterial use. Throughout this paper, the standard convention will be used referring to antibacterials as a group of chemicals comprising both the antibiotics (produced by microorganisms) and anti-microbials (of synthetic origin). The emphasis will be on systemic antibacterials used as drugs rather than those chemicals used topically or primarily used for disinfection of surfaces or the culture water (e.g. quaternary ammonium compounds, chloramine-T, iodophores).

6.2   ANTIBACTERIALS USED IN AQUACULTURE

Most systemic antibacterials employed in aquaculture generally fall within eight categories (Alderman 1988; Meyer & Schnick 1988; Alderman & Michel 1992). These categories, along with some of the representative drugs, are:

Antibiotics:
- Tetracyclines – oxytetracycline, chlortetracycline, and doxycycline.
- Macrolide antibiotics –erythromycin
- β-lactam antibiotics – amoxycillin, penicillin
- Aminoglycosides – streptomycin, neomycin, kanamycin
- Chloramphenicol

Antimicrobials:

- Sulphonamides – sulphamerazine, sulphadiazine, sulphamethylphenazole, sulphisoxazole, sulphathiazole, sulphamonomethoxine, sulphadimethoxine, and sulphanilimide. Sometimes formulated with a potentiator such as trimethoprim or ormetoprim.
- Quinolones – nalidixic acid, piromidic acid, oxolinic acid and flumequine
- Nitrofurans – nitrofurazone, furazolidone, nifurpirinol and furaltadone.

Of the many antibacterials used in agriculture or human medicine, only a small fraction has been tested for potential applications in aquaculture, and a smaller fraction still (probably less than 24) is routinely used in some segment of the world-wide aquaculture industry. The antibacterials that are used in any given industry segment depend upon the species cultured, the principal pathogens, drug availability, historical convention, and approval by regulatory authorities for safety and efficacy. The list of approved antibacterials varies widely among countries (Table 6.1). The United States is one of the most restrictive countries in approving aquacultural antibacterials, with only three antibacterials that are generally approved for use on food fish. This extremely limited list is probably due to both the relatively small size of the aquaculture industry in the country, as well as the rigorous approval process required by the US Food and Drug Administration (FDA) for drugs of all types and uses. One of the three approved drugs, sulphamerazine, is no longer commercially available.

Japan, with 26 approved antibacterials for aquaculture, has one of the most extensive lists, probably because of the size and long history of its aquaculture industry. This number of antibacterials may be unnecessary to support the aquaculture industries in most countries, and it is doubtful whether it would be commercially practical to obtain the extensive information that typically would be needed for product approval of so many compounds. However, restricting antibacterial chemotherapy to only one or two compounds may not be the best course of action either, as it forces excessive reliance upon a few drugs, and increases the selective pressure for antibacterial resistance.

In many developing countries there is no established procedure for the approval of aquacultural antibacterials. In Asia, for example, there is no regulation of antibacterial use in Bangladesh, India, Indonesia, Nepal, Pakistan and Sri Lanka (ADB/NACA 1991). In these countries, the antibacterials used are limited only by what the grower can acquire and their potentially prohibitive costs. Reliable information on appropriate treatment dose or withdrawal times may be lacking entirely and left only to the judgment of the individual user.

Given the current global market place, it is now common for aquaculture products to be marketed and consumed in countries other than the one in which they were produced. This trade raises the peculiar situation of aquaculture products grown using certain antibacterials, being sold in countries that would prohibit their own producers from using the very same drugs. For example, nifurazolidone was heavily

Table 6.1 Antibacterials used in aquaculture of food species in selected countries.

	Canada	Japan	Phillipines	Norway	United States
Species	aquaculture in general, but mostly salmonids[1]	primarily yellowtail[2]	penaeid shrimp	salmonids	salmonids[3]
Reference	E.A. Black, pers. comm.	Schnick 1991, ADB/ NACA 1991, Okamoto 1992	Primavera, et al. 1993	Norwegian Medicinal Depot, unpublished data	Schnick 1991
Antibacterials	erythromycin oxolinic acid oxytetracycline penicillin G sulphadimethoxine and ormetoprim sulphamerazine	amoxycillin ampicillin chlortetracycline colistin[2] doxycycline erythromycin flumequine florfenicol josamycin kitasamycin lincomycin miloxacin nalidixic acid[2] nifurstylenic acid novobiocin oleandomycin oxolinic acid oxytetracycline piromidic acid[2] sodium polystyrene sulphonate spiramycin sulphadimethoxine[2] sulphamonomethoxine sulphamonomethoxine and ormetoprim[2] sulphisoxazole tetracycline thiamphenicol	chloramphenicol[4] doxycycline[5] erythromycin[5] furazolidone[5] oxytetracycline	florfenicol flumequine furazolidone oxolinic acid oxytetracycline sulphadiazine and trimethoprim	oxytetracycline sulphamerazine[6] sulphadimethoxine and ormetoprim

[1] List includes therapeutants both generally authorised (only oxytetracycline) and available with an emergency-release permit.

[2] Aquacultural therapeutants are approved in Japan on a species-specific basis. Except for those footnoted, the drugs listed are approved for yellowtail.

[3] Therapeutants listed are approved for salmonids. Only oxytetracycline is approved for catfish, striped bass and lobsters. This list does not include those compounds for which an organisation has been given approval for experimental testing of a potential candidate antibacterial (e.g. amoxycillin, erythromycin).

[4] In 1990 the Phlippine government banned chloramphenicol for use on animals destined for human consumption (ADB/NACA 1991), but a survey conducted in the same year (Primavera *et al.* 1993) documented its use at 4 out of 21 farms surveyed.

[5] Available for use in Philippine prawn culture, but not widely used.

[6] Sulphamerazine is approved for use, but is no longer commercially available.

used by Norwegian salmon producers in the late 1980s and was used in small quantities through the early 1990s. Some of these fish are marketed in the United States, although in that country the FDA has refused to approve the use of nitro-furans by US salmon producers because of concerns over potential carcinogenicity and mutagenicity. Moreover, Norwegian fish imported into the US are not analysed for residues of nitrofurans because of analytical obstacles, including rapid biotransformation (K. Greenlees & A. Montgomery, US FDA, pers. comm.).

A second prominent example is the use of chloramphenicol. This drug can be toxic in very low concentrations, posing potential risk to farm workers and consumers of contaminated seafood. It also has critical medical uses (e.g. drug of choice against typhoid) that could be compromised by elevated levels of antibacterial resistance. For these reasons, use of chloramphenicol on aquaculture animals intended for human consumption is either not approved (US), explicitly banned (Canada, Denmark), or heavily restricted (United Kingdom) in many countries. It is, however, used in aquaculture in some European countries, including France (Alderman *et al.* 1994), and is heavily used in prawn and shrimp culture in Ecuador and Asia (Brown 1989; ADB/NACA 1991).

6.3   PATTERNS OF ANTIBACTERIAL USE

Although some systemic antibacterials can attain therapeutic concentrations in tissue by bath treatment (e.g. nifurpirinol), most are administered to the cultured animal as a feed supplement. Most often, antibacterials are administered therapeutically, i.e. to treat an ongoing disease outbreak. Under conditions of therapeutic use, antibacterials are administered only sporadically. In salmonid net-cage culture in the north-western United States, for example, antibacterial use for 30 days (not usually continuous) out of the year might be considered typical (Weston 1994). Prophylactic application, i.e. treating when there is a risk of infection in order to prevent disease, is sometimes done, but is less common and more controversial. Some salmon growers will use antibacterials prophylactically at the time of transition from fresh to salt water, when the risk of disease is greatest. Some shrimp and bivalve hatcheries will also maintain low levels of antibacterials in the culture water to ward off disease (Brown 1989).

The amounts of antibacterial used in a given culture facility can be extremely variable, depending upon water temperature, culture conditions (especially stocking density), site characteristics, propensity of the operator to use chemotherapeutants, and many other factors. Figure 6.1 illustrates the potential variation among farms, based on data from four farms in Puget Sound, Washington, USA (Weston 1994). All farms were net-cage facilities, culturing Atlantic salmon (*Salmo salar*) to a market size of 3–6 kg, and producing 300–700 t per year. All farms treated largely with oxytetracycline, with only minor use of sulphadimethoxine/ormetoprim and amoxycillin. In any given year, the amounts of antibacterial used (weight of active ingredients only) at a given farm varied by a factor of two or more.

Differences between farms were even more striking. The farm using the least antibacterial (Farm C) required only 0–63 kg antibacterial/year, or on a production-weighted basis, 0 to 0.15 kg/t fish produced. In contrast, the heaviest user (Farm A) treated with 224–447 kg of antibacterial drugs/year (0.8–3.4 kg/t). The average amount of antibacterial agent needed to produce 1 t of salmon was 40 times greater at

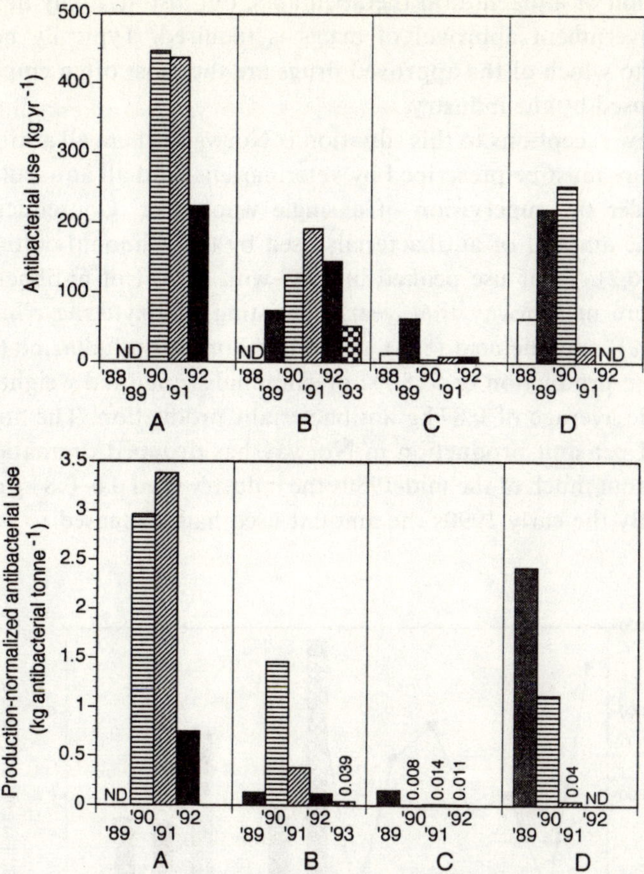

Fig. 6.1 Antibacterial use at four salmon net-cage farms in Puget Sound, Washington, USA. The upper panel shows the weight of active ingredients (largely oxytetracycline) used at each farm over 3–5 years. The lower panel presents these data normalised to farm production at the grow-out facility, but does not include any antibacterials used in the hatchery stage. From Weston (1994).

Farm A than at Farm C. This high use of antibacterials at Farm A was due to a less desirable site, a high proportion of fish unvaccinated for vibriosis, and a perceived need to treat the entire farm prophylactically as water temperatures rose in the summer.

Collecting data on the quantities of antibacterials used, as presented above, required approaching each culturist individually and obtaining their voluntary release of husbandry information. This situation is usual, since in the majority of countries using antibacterials in aquaculture diseases are diagnosed by the grower rather than a trained veterinarian, no prescription system for distribution of the drugs exists, and there are no centralised government records on the amount of antibacterials used. Certainly this situation occurs in most developing countries lacking any governmental regulation of aquacultural therapeutants, but also in many developed countries where government approval of drugs is required. Typically no records are maintained as to which of the approved drugs are the most often employed or what quantities are used by the industry.

One of the few exceptions to this situation is Norway, where all antibacterials used in salmon culture must be prescribed by veterinarians, and all aquaculture drugs are distributed under the supervision of a single wholesaler. Consequently, data are available on the amount of antibacterials used by the salmonid culture industry in Norway (Fig. 6.2). Their use peaked in 1987 with 48.57 t of antibacterials used in salmonid culture in Norway that year, consisting of oxytetracycline (56%), furazolidone (33%), oxolinic acid (8%), and trimethoprim/sulfadiazine (4%). Given a 1987 Norwegian production of 55 800 t of salmonids (ungutted weight), this equates to a nationwide average of 0.87 kg antibacterial/t production. The amount of antibacterials used per unit production in Norway has dropped dramatically in recent years. Throughout much of the mid-1980s the industry used 0.6–0.8 kg antibacterials/t production. By the early 1990s the amount used had decreased to 0.2 kg/t, and in

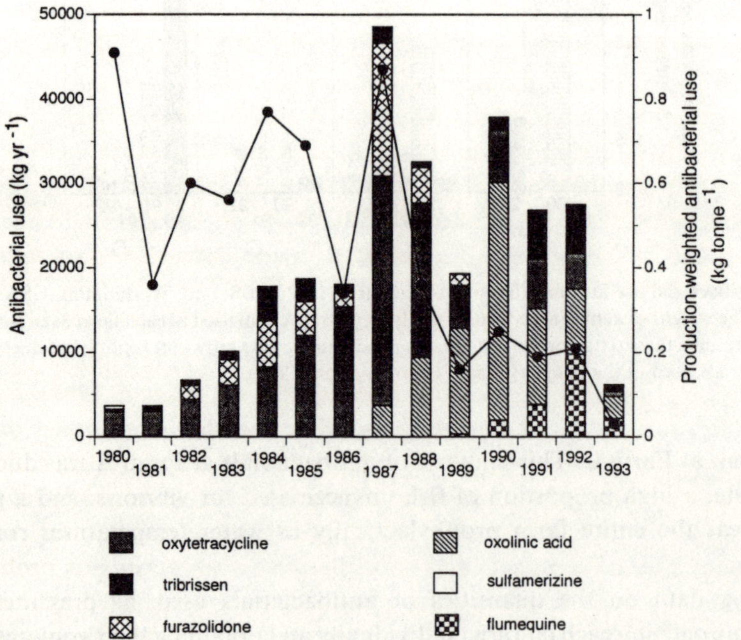

Fig. 6.2 The amounts of antibacterials used in Norway annually for the production of Atlantic salmon and rainbow trout. The weight of active ingredients (left scale) is represented by the stacked histograms; the amount of antibacterials per unit production (right scale) is represented by the line. The scale does not show use of small quantities (< 150 kg) of Tribrissen in 1989, furazolidone in 1990, 1991, and 1993, and florfenicol in 1993. Data from Grave *et al.* (1990) and pers. comm. to A. Bergheim from the Norwegian Medicinal Depot and the Norwegian Fish Farmers Association.

1993 use was 0.03 kg/t. This reduction may, in part, reflect improved husbandry practices and greater use of vaccination, but it should also be recognised that the current drugs of choice (oxolinic acid and flumequine) require much smaller doses per treatment than oxytetracycline, the historical dominant. Thus, the weight of antibacterials used would be much smaller even if the number of treatments remained the same.

6.4   PERSISTENCE OF DRUG RESIDUES IN THE ENVIRONMENT

Environmental fate and effects of antibacterials are potential concerns only because a substantial fraction of the medication provided either does not reach or remain in the cultured animal, and in many types of culture antibacterials can be transported beyond the boundaries of the culture facility. Since most antibacterials are administered via the feed, one of the first routes of loss is waste feed that is not ingested by the cultured animal. Amount of feed waste is likely to be highly farm-specific and estimates vary widely. It has been estimated that 1–5% of the feed is wasted in land-based salmonid culture (VKI, 1976), and 15–40% in salmonid net-cage culture (Gowen *et al*. 1985; Rosenthal *et al*. 1988; Thorpe *et al*. 1990). Since ingestion rates are generally lower among diseased fish because of poor appetite or reduced palatability of the diet, it is reasonable to assume that waste of antibacterial-medicated feed is at the high end of these estimates or even greater.

Of the portion of antibacterial consumed by the cultured animal, a substantial fraction may pass into the environment unaltered, through either the faeces or the urine. Oxytetracycline, one of the most widely used antibacterials in aquaculture world-wide, is notorious in this regard. The vast majority of oxytetracycline supplied in medicated feed can be found in the feces in the parent form as indicated by relatively low tissue oxytetracycline concentrations attained after medication (Björklund & Bylund 1990; Rogstad *et al*. 1991), oxytetracycline presence in hatchery effluent at concentrations that account for nearly all of the drug supplied (Smith *et al*. 1994a), and direct measurement of gut absorption (Cravedi *et al*. 1987). It has been estimated that only 7–9% of the oxytetracycline ingested is absorbed during gut passage in freshwater rainbow trout (Cravedi *et al*. 1987). Since divalent cations in sea water are likely to reduce oxytetracycline bioavailability (Neuvonen 1976; Lunestad & Goksøyr 1990), intestinal absorption in marine organisms may be even less than the 7–9% estimate. Given feed waste and poor digestive absorption, it is probable that greater than 95% of the oxytetracycline provided is not assimilated by the cultured organism, has no therapeutic value, and leaves the farm via the effluent.

All antibacterials administered as feed additives can reach the environment through waste feed, but the importance of fecal sources depends on the particular drug. Very little chloramphenicol (< 1% of drug ingested) is released via the feces. Fecal sources are more important for oxolinic acid, where 62–86% of the ingested drug remains in the feces (rainbow trout, Cravedi *et al*. 1987).

The selective pressure for development of antibacterial resistance in microbes (see [6.6]) is dependent upon the duration of drug exposure. Since most culturists deal with antibacterials only intermittently, this treatment schedule would tend to minimise the potential for development of resistance. However, if antibacterial residues remain in the environment surrounding the culture facility for long periods of time after chemotherapy, these residues would reduce any advantage gained by intermittent application. Antibacterial residues in the surrounding water are not a major concern given the rapid dilution and susceptibility of some drugs to photodegradation, e.g. oxytetracycline (Oka *et al.* 1989; Samuelsen 1989) and furazolidone (Samuelsen *et al.* 1991).

However, the sediments serve as a long-term reservoir for residues of many drugs (Table 6.2). It should be recognised that these data are derived exclusively from marine waters and that the field data are entirely from salmonid net-cage culture systems. Very few data are available from fresh water. It should also be noted that the studies shown in this table used very different methods to measure persistence of residues (e.g. flow-through vs static systems; residues homogenised into the sediment vs added as a superficial layer) so some discrepancies between results should be expected.

Of the antibacterials examined, oxytetracycline is among the most persistent. The drug does not appear to be microbially degraded (Brander & Pugh 1977; Robins-Brown *et al.* 1979), and is lost from the sediment only by dissolution and diffusion into the overlying water (Samuelsen 1989). Under conditions of rapid sedimentation, such as would be expected near many aquaculture facilities, sediment containing oxytetracycline residues may be quickly buried, and the drug may persist indefinitely. Fig. 6.3 shows data collected under a salmon net-cage farm before, during and after medication with 186 kg of oxytetracycline (Capone *et al.*, 1994a). Oxytetracycline concentrations of 1–4 mg/kg were observed in superficial sediments (0–2 cm) shortly after treatment, with concentrations decreasing to less than 0.5 mg/kg after 2 months as oxytetracycline diffused into the overlying water and the contaminated sediments were diluted with unmedicated feed and feces and advected sediments from surrounding areas.

It is particularly noteworthy that about 1 mg/kg oxytetracycline was present in superficial sediments before treatment and in sediments buried 2–4 cm below the surface. There had been no use of oxytetracycline at the farm since the previous summer, indicating that residues in these buried sediments had persisted for at least 10 months and possibly longer. Results indicating an extremely long persistence of the drug are not unusual. Persistence of 6 months to a year or more is commonly reported (Table 6.2). Concentrations of < 10 mg/kg are commonly reported under salmonid net-cages (Jacobsen & Berglind 1988; Björklund *et al.* 1990; Björklund *et al.* 1991; Coyne *et al.* 1994), although concentrations as high as 490 mg/kg have been noted (Samuelsen *et al.* 1992a). In marine systems, however, only about 5% of the oxytetracycline is in a form retaining antibacterial activity, and the remainder is

Table 6.2 Persistence of antibacterial residues in sediments. Bold text indicates data from sediments beneath a farm; all other data from laboratory microcosms.

Antibacterial	Half-life (days)	Persistence[1] (days)	Farm or laboratory data	Reference
Flumequine	155	> 185	L	Hansen *et al.* 1992
				Samuelsen *et al.* 1994
Furazolidone	0.75	9	L	Samuelsen *et al.* 1991
Ormetoprim		> 2– < 23	L	D.G. Capone & D.P. Weston, unpublished data
		< 30	L	Samuelsen *et al.* 1994
Oxolinic acid		**6**	**F**	Björklund *et al.* 1991
		> 77	L	Björklund *et al.* 1991
	165	> 185	L	Hansen *et al.* 1992
	48	> 210	L	Samuelsen 1992
		> 180	L	Samuelsen *et al.* 1994
Oxytetracycline	**9–419**	**>7– > 308**	**F**	Björklund *et al.* 1990
		> 12	**F**	Björklund *et al.* 1991
		> 77	L	Björklund *et al.* 1991
		> 300	**F**	D.G. Capone & D.P. Weston, unpublished data
	16–34	> 60	L	D.G. Capone & D.P. Weston, unpublished data
	16	**> 33– < 71**	**F**	Coyne *et al.* 1994
	125	> 185	L	Hansen *et al.* 1992
		> 84	**F**	Jacobsen & Berglind 1988
	70	> 84	L	Jacobsen & Berglind 1988
	32	**> 39**	**F**	Samuelsen 1989
	30–64	> 220	L	Samuelsen 1989
	55	> 210	L	Samuelsen 1992
	87–144	**> 550**	**F**	Samuelsen *et al.* 1992a
		> 180	L	Samuelsen *et al.* 1994
Sulphadiazine		> 180	L	Samuelsen *et al.* 1994
Sulphadimethoxine		> 2– > 22	L	D.G. Capone & D.P. Weston, unpublished data
		> 180	L	Samuelsen *et al.* 1994
Trimethoprim		> 30– < 60	L	Samuelsen *et al.* 1994

[1] Length of time after medication during which detectable residues were present in the sediment.

complexed with magnesium, calcium and other divalent cations (Lunestad & Goksoyr 1990).

The quinolones oxolinic acid and flumequine are also very persistent in sediments, and detectable residues may remain for many months after drug treatment. Like oxytetracycline, the drugs form complexes with cations. Most investigators have found little or no loss of antibacterial activity of flumequine and oxolinic acid residues in sediments even after 6 months (Hansen *et al.* 1992; Samuelsen *et al.* 1994).

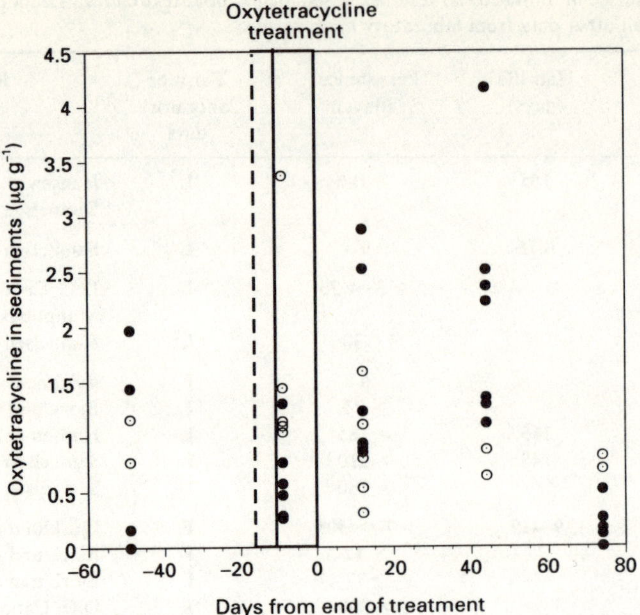

Fig. 6.3   The concentration of oxytetracycline in sediments beneath a salmon net-cage facility in Puget Sound, Washington, USA. Samples were analysed from two horizons within the sediment column: 0–2 cm (dark circles) and 2–4 cm (open circles). The vertical lines indicate the period of oxytetracycline usage at the farm, with a few cages receiving medicated feed beginning at the first dashed line and the entire farm receiving chemotherapy at the first solid line.

Björklund *et al.* (1991) found loss of oxolinic acid's antibacterial activity in the sediment after 10 days, but it is not known whether this reflects a bioavailability phenomenon, increased microbial resistance, or a reduction below minimum inhibitory concentration.

Furazolidone is degraded microbially very rapidly in sediments, with a half-life of less than a day. The principal degradation product has no antibacterial activity (Samuelsen *et al.* 1991).

Existing data are inconclusive with respect to the sulphonamides (sulphadiazine and sulphadimethoxine) with very different persistences reported, probably reflecting differences in experimental conditions, e.g. flow-through in Capone *et al.* (1994b) and a sealed vessel in Samuelsen *et al.* (1994). The potentiators with which sulphonamides are often administered, ormetoprim and trimethoprim, persist for no more than 2 months in marine sediments, and possibly much less.

6.5   DRUG RESIDUES IN NON-TARGET ORGANISMS

In those countries in which antibacterial use in aquaculture is regulated, culturists are required to hold the animals for a specified withdrawal period following cessation of antibacterial use in order to allow time for elimination of drug residues from edible tissues (see [6.8]). When withdrawal schedules are followed, this practice provides substantial protection to the consumer against incidental ingestion of antibacterials. However, no comparable protection would be afforded consumers of wild fish and invertebrates collected in the vicinity of an aquaculture operation if antibacterials reached these non-target animals. This concern is particularly relevant to polyculture, such as the production of shellfish in areas downcurrent of fish culture where antibacterials may be used.

Aquaculture operations are often characterised by an increased density of wild fish and invertebrates in the vicinity. Wild fish are abundant near net-cages (Loyacano & Smith 1976; Beveridge 1984) and crabs and starfish have been found in elevated abundance beneath mussel rafts (Romero *et al.* 1982; López-Jamar *et al.* 1984). These organisms are attracted for a variety of reasons, including the shelter provided by the culture structure, waste feed, encrusting organisms on the structures, and the dense assemblage of macrofauna characteristic of organically-enriched sediments (Beveridge 1984; Carss 1990; Weston 1990). Because of the enhanced densities of wild fish and invertebrates, the area surrounding some aquaculture operations may be frequented by sport and commercial fishermen.

Only a few studies have been done, and these have been limited to net-cage culture, but the data indicate that cultured bivalves or wild fish and invertebrates in the vicinity of a fish farm can accumulate antibacterials in their tissue to levels which would be considered unacceptable for human consumption. In an extensive study surrounding salmonid net-cage farms treating with oxolinic acid, Samuelsen, *et al.* (1992b) reported oxolinic acid residues in wild saithe, mackerel, cod, pollack, ballan wrasse, salmon, flounder, cancrid crabs and mussels with residues persisting for 1–2 weeks after cessation of chemotherapy. The maximum concentration found in edible tissues was 12.5 µ/g in the muscle of a saithe, whereas Norwegian authorities would allow no measurable oxolinic acid (< 0.1 µg/g) in cultured fish intended for human consumption. A later study (Ervik *et al.* 1994) produced similar results, with measurable oxolinic acid or flumequine residues in 84% of the 189 saithe tested. Oxolinic acid concentrations in fish tissue exceeded 0.48 µ/g in 50% of the fish and reached a maximum of 15.7 µg/g.

Similar reports have come from farms treating with oxytetracycline, including the presence of residues in wild bleak, roach and mussels (Björklund *et al.* 1991; Møster 1986). Oysters are able to bioaccumulate oxytetracycline when exposed in the laboratory to unrealistically high concentrations of the drug in dissolved form or in medicated feed (Black *et al.* 1991), but our own studies have failed to find residues in oysters placed directly beneath net-pens during and after treatment with nearly 200 kg oxytetracycline (Capone *et al.* 1994b). We, however, did find residues of up to 3.8 µg/g oxytetracycline in the back fin meat of red rock crabs (*Cancer productus*) collected under the farm 12 days after the end of treatment. This concentration exceeded by about 40-fold the US FDA allowance of 0.1 µg/g in edible flesh. This result is of particular concern because employees at this farm caught crabs on site for their personal consumption.

6.6 STIMULATION OF ANTIBACTERIAL RESISTANCE

The use of antibacterials in aquaculture is often accompanied by an increase in the proportion of antibacterial-resistant bacteria in the surrounding environment. Bacteria may acquire resistance by a variety of mechanisms, including blocking entry of the drug into the cells, inactivating the drug, altering affinity of the target site for the drug, or reducing the cell's dependence on the blocked pathway (Davies & Smith 1978). The emergence of antibacterial resistance in fish pathogens represents a very real threat to the aquaculture industry particularly given the limited number of antibacterials available in many countries. Many investigators have reported a disturbing pattern of increases in resistance in pathogens following adoption and widespread use of an antibacterial by the aquaculture industry (Aoki *et al.* 1983; Hastings & McKay 1987; Aoki *et al.* 1990; Meier *et al.* 1992; Richards *et al.* 1992).

In one of the first studies to consider environmental consequences of this resistance, Austin (1985) found that use of oxolinic acid, oxytetracycline, or a potentiated sulphonamide at rainbow trout farms all increased the proportion of bacteria in the effluent resistant to these drugs. Many other studies in the vicinity of salmonid netcages have reported elevated levels of oxytetracycline resistance in the bacteria of sediments near the farms (Torsvik *et al.* 1988; Björklund *et al.* 1991; Husevåg *et al.* 1991; Samuelsen *et al.* 1992a; Kerry *et al.* 1994, 1995).

An increased antibacterial resistance in sedimentary bacteria is often the most sensitive environmental indicator of past antibacterial use. Elevated resistance may be observed in conjunction with reductions in total microbial density or changes in functional properties of the sedimentary microbial community such as sulphate reduction rates (Hansen *et al.* 1992), but levels of antibacterial resistance may increase even in the absence of these other effects (Hsu *et al.* 1992; Samuelsen *et al.* 1992a; Weston *et al.* 1994).

The concept of antibacterial resistance in aquaculture, the methods of quantifying it, and the many unknowns concerning mechanisms by which resistance is developed and maintained have been thoroughly reviewed by Smith *et al.* (1994). These authors identified a number of limitations in current understanding of antibacterial resistance and difficulties in interpreting resistance data from environmental samples.

First, the use of one antibacterial agent can increase levels of resistance not only to that specific drug but to many others (Wood *et al.* 1986), even those using very different modes of antibacterial action. This cross-resistance may be attributable to the production of plasmids encoding resistance to multiple antibacterials (Brazil *et al.* 1986), but in some cases is attributable to mechanisms of resistance that are not plasmid-mediated (Smith *et al.* 1994b).

As an example of cross-resistance, dosing of sediments in laboratory microcosms with oxytetracycline increased the proportion of oxytetracycline-resistant bacteria from < 1% before treatment to 17–33% immediately after treatment (Fig. 6.4). However, the proportion of bacteria resistant to the potentiated sulphonamide Romet® 30 increased even more substantially, attaining 100% resistance in one case, even though these sediments were exposed only to oxytetracycline. Cross-resistance between oxytetracycline and Romet® 30 was also observed at a salmon net-cage farm in response to treatment only with oxytetracycline (Fig. 6.4). Other investigators have also frequently reported cross-resistance between other aquacultural antibacterials, such as between oxytetracycline and oxolinic acid (Nygaard *et al.* 1992; Hansen *et al.* 1992; Ervik *et al.* 1994).

A second difficulty is that antibacterial resistance does not always respond in a predictable fashion correlating with the amount of drugs used or with the concentrations of residues in the environment. In studies at two Irish net-cage farms, sediments beneath both farms contained about 10 mg/kg oxytetracycline following

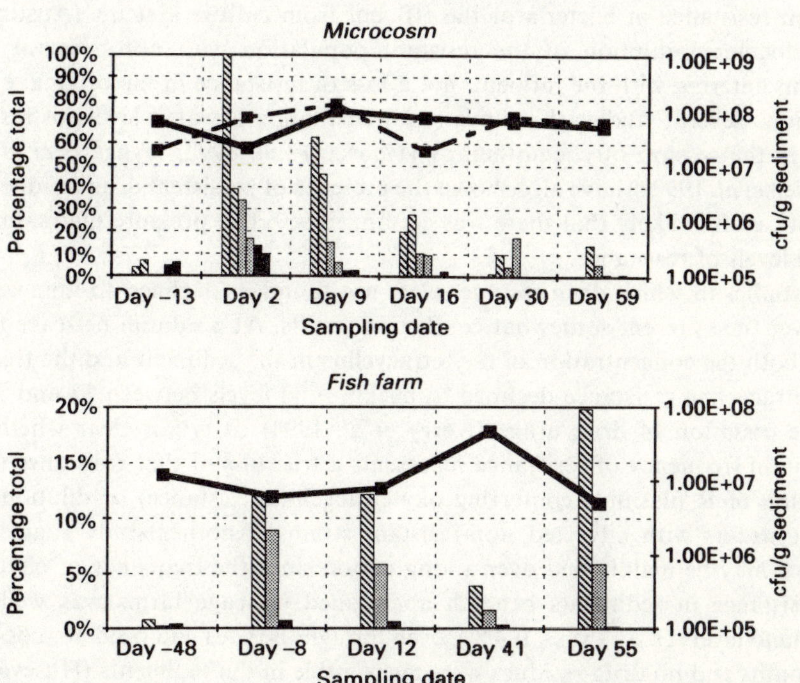

Fig. 6.4 The effect of the addition of oxytetracycline-medicated feed on sedimentary bacteria in laboratory microcosms and at a farm site in Puget Sound, Washington, USA. The left histogram in each set (or paired treatments in the case of the microcosms) represents the proportion of microbes resistant to the sulphonamide Romet®30. The middle and right histograms represent the proportions resistant to oxytetracycline and amoxycillin, respectively. Resistance was quantified as the number of colony-forming units (cfu) growing on media containing the drug as a percentage of cfu on unmedicated media (shown as solid or dashed lines). Sampling dates are expressed as the days since the end of antibacterial treatment that began on Day 10 in microcosms and Day 17 at the farm). From Gray & Herwig 1994; Gray *et al.* 1994.

chemotherapy, but whereas at one farm the frequency of oxyteteracycline resistance increased from about 1% before therapy to 16% after treatment, at the other farm resistance levels remained essentially unchanged (Kerry *et al.* 1994). In another recently completed study (Gray *et al.* 1994), the use of oxytetracycline at a salmon net-cage farm increased the proportion of culturable bacteria resistant to oxytetracycline from less than 1% to 8%, and the proportion resistant to Romet® 30 from about 1% to 12% (Fig. 6.4). Resistance levels to both drugs gradually declined over the next 2 months, but then abruptly increased to even greater levels than during chemotherapy, despite the fact that the farm had not used any additional antibacterials. Results such as this may be due to the fact that exposure to heavy metals and other toxicants can confer resistance to some antibacterials as well (Timoney & Port 1982; Baya *et al.* 1986).

A third difficulty in assessing environmental effects is that existing data are not adequate to establish how long bacteria maintain antibacterial resistance in the absence of continued selective pressure for that resistance. Studies showing a rapid decline in resistance in bacteria of the effluent from culture systems (Austin 1985) merely document dilution of the resistant population with non-resistant micro-organisms entering with the influent, not a loss of resistance in the original exposed population. Several studies that have reported finding elevated levels of resistance long after the cession of chemotherapy (Hansen *et al.* 1992; Nygaard *et al.* 1992; Samuelsen *et al.* 1992a) have also shown the presence of persistent drug residues in the sediments, so it is likely that there was continued selective pressure maintaining the elevated levels of resistance.

Two studies in which drug residues were not found or declined to immeasurable levels over time present somewhat conflicting results. At a salmon net-cage farm in Ireland both the concentration of oxytetracycline in the sediment and the frequency of oxytetracycline resistance declined to background levels between 33 and 73 days after the cessation of drug usage (Kerry *et al.* 1994). It is not clear whether this reduction in frequency of resistance represents a true loss of that resistance (e.g. no production of R plasmids conferring oxytetracycline resistance) or dilution of the resistant strains with advected non-resistant strains. Another study suggests that resistance may be maintained over a long period since the frequency of oxytetracy-cline resistance in sediments beneath abandoned net-cage farms was well above background levels (1–8% vs < 0.4%), even though the sites had been abandoned for 6–12 months and no drug residues were measurable in the sediments (Husevåg *et al.* 1991).

6.7   EFFECTS ON BIOGEOCHEMICAL PROCESSES

Deposition of feed and faeces from aquaculture enhances the density of many anaerobic and some aerobic functional groups of microorganisms (Ram *et al.* 1981, 1982). The organic-rich sediments typical in the vicinity of aquaculture operations are characterised by intense microbial activity as evidenced by high rates of oxygen consumption and sulphate reduction (Dahlback & Gunnarsson 1981; Hall & Holby 1986; Holmer & Kristensen 1992). Since antibacterial residues can be found in sediments surrounding aquaculture operations and persist there for a year or more in the case of some antibacterials, the question arises as to what effect these residues may have on natural microbial communities in the sediments. Antibacterials in cattle feed can sometimes lead to the production of a manure that is less biodegradable (Elmund *et al*. 1971). If antibacterials were to affect sedimentary microbes, this could have ramifications, positive or negative, to the culturist; e.g. microbial reduction of sulphate is responsible for the production of hydrogen sulphide with adverse effects on growth and survival of the cultured species. Antibacterial-induced reduction in microbial densities could also have indirect consequences to meio-and macrofaunal invertebrates, for which sediment microbes may be an important food source.

Despite the potential environmental effects, extremely little work has been done on changes in sedimentary microbial abundance or biogeochemical processes following aquacultural use of antibacterials. One very preliminary study (Samuelsen *et al*. 1988) found a reduction in microbial density at a farm site after oxytetracycline treatment, but no similar response was observed in laboratory microcosms. Jacobsen & Berglind (1988) found that oxytetracycline, furazolidone and trimethoprim/sulphadiazine resulted in a higher sulphide ion activity in the sediments, but interactions with pH and lack of other data on other microbial or chemical parameters make these results uninterpretable. Moreover, the study used antibacterial concentrations orders-of-magnitude greater than normally reported at aquaculture sites.

The most extensive data on biogeochemical effects of antibacterial treatment comes from work just completed at a salmon net-cage farm in the United States and in laboratory microcosms (Capone *et al*. 1994a). Use of oxytetracycline at a farm and the presence of residues of the drug in the sediments (1–4 mg/kg) had no measurable effect on total microbial density in the farm sediments, the flux of ammonium from the sediments, sulphate reduction rates, or sediment oxygen consumption. In the same study, dosing of microcosms with oxytetracycline and sulphadimethoxine/ ormetoprim at a rate intended to simulate field conditions also had no effect on the same parameters (Fig. 6.5). The lack of a response in microbial density or activity was probably due, in part, to an observed increase in antibacterial resistance that compensated for the loss of more susceptible micro-organisms. Other mitigating factors included complexation of the oxytetracycline with divalent cations, and a rapid disappearance of the sulphonamide, measurable on the second day after treatment but below detection limit by day 22.

In the only other study on the topic, dosing of sediments with either oxytetracycline, oxolinic acid or flumequine resulted in a 40–50% reduction in microbial density and a > 90% decrease in the rate of sulphate reduction in the sediments (Hansen *et al*. 1992). The disparity between these results and the lack of microbial response observed by Capone *et al*. (1994a) is probably due to differences in the concentrations of

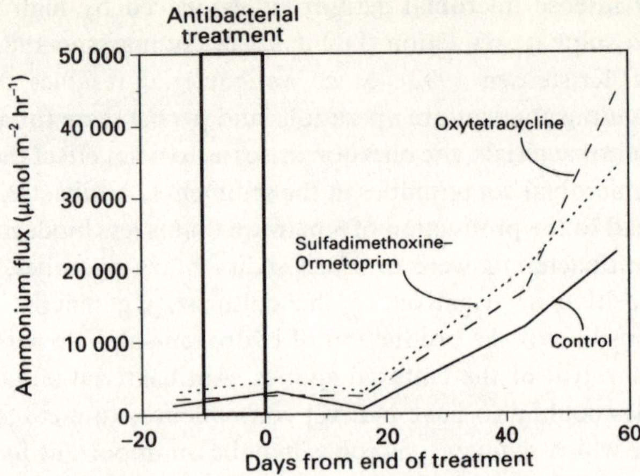

Fig. 6.5 Flux of ammonium from the sediments to the overlying water in microcosms following: (1) control treatment with unmedicated feed only; (2) treatment with oxytetracycline-medicated feed; and (3) treatment with Romet® 30-medicated feed containing sulphadimethoxine and ormetoprim. Each line represents the mean of two duplicate treatments. From Capone *et al.* (1994b).

antibacterials used. On the basis of field data from Norwegian net-cage sites, Hansen *et al.* (1992) dosed the sediments with 400 mg/kg oxytetracycline and 100 mg/kg of the other drugs. However, Capone *et al.* (1994a) used 5–15 mg/kg oxytetracycline in the microcosms and reported < 4 mg/kg in the farm sediments, concentrations more typical of most farm sites studied (Jacobsen & Berglind 1988; Björklund *et al.* 1991; Coyne *et al.* 1994).

6.8   HUMAN HEALTH ISSUES

The use of antibacterials in the culture of animals intended for human consumption raises the concern of antibacterial residues in the human diet, particularly since many antibacterials are not thermally degraded under normal cooking conditions (Moats 1988; Kitts *et al.* 1992). Unintended ingestion of antibacterials can have a number of adverse consequences, and the concerns generally fall within three categories as summarised by Yndestad (1992):

(1) *Toxicity*. Michel (1986) lists toxic effects of many antibacterials including potential carcinogenicity of the nitrofurans, and digestive or hepato-renal disorders associated with furazolidone, oxytetracycline, and sulphonamides. Of greatest concern is chloramphenicol as a causative agent for aplastic anaemia. The condition is often fatal, about 1 in 40 000 persons are predisposed to the condition, and toxicity appears to be independent of the dose (Yndestad 1992).

(2) *Hypersensitivity*. Exposure to antibacterials can evoke an allergic reaction in individuals who are presensitised to the drug. Penicillins are the best known in this regard although other antibacterials can induce allergies as well. The general

consensus, however, is that ingestion of foods containing drug residues is unlikely to be of major health significance due to hypersensitivity (FAO/WHO 1990).

(3) *Microbial resistance.* Consumption of antibacterial residues via aquaculture products could potentially exert a selective pressure for antibacterial resistance in gut microflora, including species of pathogenic enteric bacteria. Sub-therapeutic exposure via the diet could therefore potentially compromise the effectiveness of later therapeutic use of antibacterials.

Ideally, antibacterials used in aquaculture should be rapidly metabolised by the cultured animal to non-toxic compounds. However, this does not happen to some of the most widely used antibacterials (e.g. oxytetracycline, oxolinic acid, trimethoprim and ormetoprim). Therefore, the best method to avoid human dietary exposure to antibacterials in aquaculture products is the use of adequate withdrawal periods. Those countries which regulate aquacultural use of antibacterials (and many developing countries do not) generally advise or mandate a defined period of time which must elapse between chemotherapy and slaughter of the animal to allow metabolism or excretion of the drug to safe levels. Norway, for example, requires extensive testing for drug residues in salmon both before and after slaughter (O. Westbye, K. Grave & T.T. Poppe, unpublished data).

Withdrawal times for any given antibacterial exhibit considerable variation among countries. Table 6.3 illustrates the required withdrawal periods for oxytetracycline among six countries, all of which produce Atlantic salmon and/or rainbow trout. Most define a withdrawal period based on water temperatures, an approach that is reasonable given the strong temperature dependence of oxytetracycline elimination (Björklund & Bylund 1990), but some countries such as the United States and

Table 6.3   Required withdrawal times following chemotherapy of Atlantic salmon and/or rainbow trout with oxytetracycline. As an illustration of the potential range, the length of the withdrawal periods that would be required at water temperatures of 8 and 12°C are shown. (Modified from J. Brackett, unpublished data).

Country	Required withdrawal time (days)	Days at 8°C	Days at 12°C
Canada	42	42	42
Finland	40 when > 10°C 60 when < 10°C	60	40
Norway	40 when > 10°C 80 when < 9°C	80	40
Sweden	30 when > 9°C 60 when < 9°C	60	30
United States	21	21	21
United Kingdom	400 degree days	50	33

Canada, do not. Thus, there can be up to a 2–4-fold difference in the required withdrawal times between countries, even for identical species in identical water temperatures.

Protecting the consumer from ingestion of antibacterial residues in imported or even domestic aquaculture products is a difficult task, in part because of analytical limitations. The US FDA, for example, has established methods for analysis of imported aquaculture products only for chloramphenicol and oxolinic acid. No tests are conducted for residues of the many other antibacterials in widespread use, although chemical methods for some other quinolones, malachite green, and oxytetracycline are now under development (A. Montgomery, pers. comm.).

The sheer size of commerce in aquaculture products also makes thorough screening for residues a nearly impossible task. In 1990 the US imported about 40 000 t of fresh salmon, about half of which was likely to be of aquaculture origin, principally from Norway, Chile, the UK and Canada (NRC 1992). Salmon are tested by the FDA only for residues of oxolinic acid, with typically 150 imported fish and 50 domestic fish tested annually. No salmon tested in recent years has had more than trace levels of the drug (A. Montgomery, pers. comm.). Similar data are available for imported shrimp. The US imported over 200 000 t of shrimp in 1989, about half of which came from China, Ecuador, Thailand and Taiwan, where aquaculture dominates production (NRC 1992). The FDA tests about 80 shrimp samples per year, and analyses these only for chloramphenicol. Chloramphenical residues have been discovered, although infrequently. In 1992–93 there were five samples (3.2% of samples tested) containing measurable amounts of chloramphenicol, three from Thailand and two from mainland China. These shipments were refused entry into the USA (A. Montgomery, pers. comm.).

In a survey of shrimp imported into the UK, neither shrimp from Thailand or Ecuador tested positive for chloramphenicol, sulphonamides, or β-lactams, although two of 27 shrimp from Thailand tested positive for tetracyclines (Brown & Higuera-Ciapara 1992). Other data suggest the frequency of residues may be greatest for aquaculture production consumed within the country of origin rather than exported. Of 1461 tiger shrimp (*Penaeus monodon*) produced in Thailand and purchased in open markets within that country, 8.4% tested positive for antibacterial residues in a microbiological test, and later chemical tests confirmed the presence of oxolinic acid and oxytetracycline (Saitanu *et al.* 1994).

Although these data are extremely limited, they suggest that producers of salmon are controlling antibacterial use to minimise residues, but a small percentage of shrimp contains residues of antibacterials. The fact that residue screening is done on such an infinitesimally small fraction of total aquaculture production indicates that, if antibacterial use is not controlled by the producer, the consumer has minimal protection from dietary exposure to antibacterials.

6.9   CONCLUSIONS

Assessment of the environmental ramifications of antibacterial use in aquaculture is hindered by the minimal data available. No records on the quantities of antibacterials used are available in most countries, and even determining which antibacterials are used often requires reliance on grey literature and the resolution of many conflicting accounts. Data on persistence of residues in the environment has only just become available for some antibacterials and is still minimal or lacking for others. The data that are available on persistence in sediments and residues in wild fauna are exclusively from salmonid net-cage culture in Scandinavia, the UK and North America, and certainly provide a biased perspective given the range of drugs in aquaculture and their many potential applications. No data are available on tropical aquaculture, a limitation that is particularly significant in view of the often minimal regulatory control on antibacterials in many tropical countries. It should be noted, however, that much tropical culture is extensive or semi-intensive, and thus likely to require less use of antibacterials than intensive systems. Available data are largely from marine systems, but it is likely that environmental fate and effects may be quite different in fresh water for drugs such as oxytetracycline and oxolinic acid.

Information on biogeochemical effects of antibacterials is almost entirely lacking. Some data are available on stimulation of resistance in micro-organisms isolated from the cultured animals, but fewer are available on resistance in sedimentary microbes or the implications of this elevated resistance in the environment surrounding the farm to the efficacy of subsequent antibacterial treatments. Thus, while I have attempted to address the principal issues, the limitations imposed by the available data must be recognised.

The pattern of antibacterial use typical of temperate, intensive fish culture is illustrated by an example from a salmon farm (Fig. 6.6). As water temperatures increased in June, the culturist observed a rapid increase in mortality rate. The cause of the mortality was not diagnosed bacteriologically, but since the fish responded favourably to sulphamerazine, the disease was assumed to be vibriosis, and full-scale treatment with this antibacterial was initiated with apparent success. In August, mortality rate again increased, and the grower again treated with sulphamerazine. On this occasion, however, treatment was unsuccessful. Further microbial studies indicated that the disease was furunculosis caused by a sulphamerazine-resistant strain of *Aeromonas salmonicida*. The grower then treated with oxytetracycline, although by this time furunculosis was no longer causing significant mortalities, and the oxytetracycline treatment was not justifiable. This example illustrates many of the common, but potentially unwise, practices in the use of antibacterials, including failure to obtain professional diagnosis, failure to test for antibacterial sensitivity, and the unnecessary use of antibacterials when natural factors had already limited spread of the disease.

Several recommendations can be made to reduce the environmental effects of

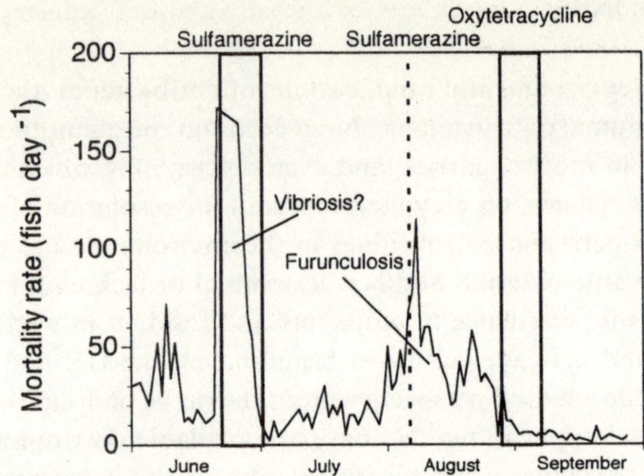

Fig. 6.6 Use of antibacterials at a Danish marine rainbow trout farm in 1978. From Christiansen & Larsen (1983). Time of second sulphamerizine treatment is approximate since this information was not provided in the original source.

antibacterial use, protect against residues in the human diet, and increase the likelihood that antibacterials will retain their therapeutic value to the culturist:

- *Improved husbandry*. Sensible husbandry practices (e.g. frequent cleaning of culture systems, rapid removal of mortalities, reduced stocking densities) can dramatically reduce the incidence of disease, and hence the need for antibacterials and other chemotherapeutants. Minimisation of feed wastage can significantly reduce release of antibacterials to the environment.
- *Vaccination*. Although only one vaccine is available for invertebrates (Alderman *et al.* 1994), fish vaccines are currently available for vibriosis (both *Vibrio anguillarum* and *V. salmonicida*), furunculosis, enteric septicaemia, and enteric redmouth. The most obvious way to avoid potential adverse consequences of antibacterial use is not to use them, and vaccines provide a partial, although incomplete, step in this direction.
- *Documentation of use*. Knowledge of the quantities of antibacterials used, both on a regional basis and at individual farms, is necessary to assess properly potential risks and identify culturists that are using drugs in amounts far out of proportion to industry norms. At the present time, however, only Norway maintains such records.
- *Diagnosis by veterinarian*. Again, salmonid culture in Norway is one of the few instances in which diseases are diagnosed by veterinarians and antibacterials dispensed only by prescription. Other segments of the aquaculture industry elsewhere are often too small to support this service, but, where feasible, professional diagnosis and prescription provides one of the best means to minimise use of antibacterials.

- *Sensitivity testing.* Many segments of the aquaculture industry have been confronted with an emerging resistance in pathogens that had long been sensitive to commonly-used antibacterials. Therefore, it is prudent to test the sensitivity of the pathogen to the intended antibacterial, typically by disc diffusion methods, prior to full-scale use. Such a practice will avoid needless mortalities while treating with an ineffective drug, and the associated economic costs to the grower. In fast-spreading epizootics, where immediate control is essential, pre-treatment sensitivity testing may not be possible, but even in these cases testing after the fact helps to document temporal trends in antibacterial resistance and provide insight into emerging patterns.

- *Avoiding consumption of wild fish and shellfish.* The limited data indicate that wild fish, crustaceans and some molluscs in the vicinity of farms using antibacterials may contain drug residues in their tissues at levels which exceed regulatory thresholds, and that these residues may persist for at least 2 weeks after the cessation of chemotherapy. As a precaution, consumption of seafood from the farm area during this period should be avoided.

- *Avoiding prophylactic use.* Long-term use of antibacterials at sub-therapeutic dosages runs the risk of encouraging proliferation of drug-resistant pathogen strains. Given the few antibacterials that are commercially available or have received regulatory approval in many countries, the industry must take steps to preserve the effectiveness of those available, and usage for non-therapeutic purposes should be avoided if at all possible.

- *Collection of fate and effects data prior to drug approval.* Data are lacking on the aquatic fate and effects of many commonly used antibacterials, and what little is known has largely become available only in the past five years. Collection of these types of data are needed prior to general approval of new antibacterials so that we may avoid the current situation of initiating fate and effects studies only after an antibacterial has been in widespread use for many years and tonnes have been already released to the aquatic environment.

ACKNOWLEDGEMENTS

I appreciate the assistance of Ms Carol Haley, Mr Kevin Greenlees and Mr Alfred Montgomery of the US Food and Drug Administration in providing data for this paper. Dr Peter Smith kindly provided several pre-publication manuscripts of the work he and his colleagues have done at farms in Ireland. The draft manuscript benefited from critical reviews of Dr Deborah Penry and two anonymous reviewers. Collection of data presented on Puget Sound net-cage culture was supported by a grant to the author from the National Oceanic and Atmospheric Administration, and used with permission of Drs Douglas Capone and Russell Herwig.

REFERENCES

ADB/NACA. (1991) Fish health management in Asia-Pacific. Report on a regional study and workshop on fish disease and fish health management. *ADB Agriculture Department Report Series*, **1**. Network of Aquaculture Centres in Asia-Pacific, Bangkok.

Alderman, D.J. (1988) Fisheries chemotherapy: a review. In: *Recent Advances in Aquaculture*, Vol. 3 (eds J.F. Muir & R.J. Roberts), pp. 1–61. Croom Helm, London.

Alderman, D.J. & Michel, C. (1992) Chemotherapy in aquaculture today. In: *Problems of Chemotherapy in Aquaculture: From Theory to Reality* (eds C.M. Michel & D.J. Alderman), pp. 3–22. Office International des Epizooties, Paris.

Alderman, D.J., Rosenthal, H., Smith, P., Stewart, J. & Weston, D. (1994) Chemicals used in mariculture. *ICES Cooperative Research Report*, **202**.

Aoki, T., Kitao, T., Iemura, N., Mitoma, Y. & Nomura, T. (1983) The susceptibility of *Aeromonas salmonicida* strains isolated in cultured and wild salmonids to various chemotherapeutics. *Bulletin of the Japanese Society of Scientific Fisheries*, **49**, 17–92.

Aoki, T., Takami, K. & Kitao, T. (1990) Drug resistance in a non-hemolytic *Streptococcus* sp. isolated from cultured yellowtail *Seriola quinqueradiata*. *Diseases of Aquatic Organisms*, **8**, 171–7.

Austin, B. (1985) Antibiotic pollution from fish farms: effects on aquatic microflora. *Microbiological Sciences*, **2**, 113–17.

Barinaga, M. (1990) Fish, money, and science in Puget Sound. *Science*, **247**, 631.

Baya, A.M., Brayton, P.R., Brown, V.L., Grime, D.J., Russek-Choen, E. & Colwell, R.R. (1986) Coincident plasmids and antimicrobial resistance in marine bacteria isolated from polluted and unpolluted Atlantic Ocean samples. *Applied and Environmental Microbiology*, **51**, 1285–92.

Beveridge, M.C.M. (1984) Cage and pen fish farming: Carrying Capacity Models and Environmental Impact. *FAO Fisheries Technical Paper*, **255**. FAO, Rome.

Björklund, H. & Bylund, G. (1990) Temperature-related absorption and excretion of oxytetracycline in rainbow trout (*Salmo gairdneri* R.). *Aquaculture*, **84**, 363–72.

Björklund, H., Bondestam, J. & Bylund G. (1990) Residues of oxytetracycline in wild fish and sediments from fish farms. *Aquaculture*, **86**, 359–67.

Björklund, H.V., Råbergh, C.M.I. & Bylund, G. (1991) Residues of oxolinic acid and oxytetracycline in fish and sediments from fish farms. *Aquaculture*, **97**, 85–96.

Black, E.A., Little, J.M., Brackett, J., Jones, T. & Iwama, G.K. (1991) Co-culture of fish and shellfish: the implications for antibiotic contamination of shellfish. ICES CM 1991/F:23.

Brander, G.C. & Pugh, D.M. (1977) *Tetracyclines. Veterinary Applied Pharmacology and Therapeutics*, 3rd edn. Lea and Febiger, Philadelphia.

Brazil, G., Curley, D., Gannon, F. & Smith, P. (1986) Persistence and acquisition of antibiotic resistance plasmids in *Aeromonas salmonicida*. In: *Antibiotic Resistance Genes: Ecology, Transfer and Expression* (eds S.B. Levy & R.P. Novick), pp. 107–12. Cold Springs Harbor Laboratory, Cold Springs Harbor, NY.

Brown, J.H. (1989) Antibiotics: their use and abuse in aquaculture. *World Aquaculture*, **20**, 34–43.

Brown, J.H. & Higuera-Ciapara, I. (1992) Antibiotic residues in farmed shrimp – A developing problem? *Problems of Chemotherapy in Aquaculture: From Theory to Reality* (eds C.M. Michel & D.J. Alderman), pp. 223–30. Office International des Epizooties, Paris.

Capone, D.G., Miller, V., Love, J. & Shoemaker, C. (1994a). Effect of aquacultural antibacterials on biogeochemical processes in sediments: field and microcosm observations. In: *Environmental Fate and Effects of Aquacultural Antibacterials in Puget Sound* (eds D.P. Weston, D.G. Capone, R.P. Herwig & J.T. Staley). Appendix 3. Report from the University of California at Berkeley to the National Oceanic and Atmospheric Administration.

Capone, D.G., Weston, D.P., Miller, V. & Shoemaker, C. (1994b). Antibacterial residues in marine sediments and invertebrates following chemotherapy in aquaculture. In: *Environmental Fate and Effects of Aquacultural Antibacterials in Puget Sound* (eds D.P. Weston, D.G. Capone, R.P. Herwig & J.T. Staley). Appendix 2. Report from the University of California at Berkeley to the National Oceanic and Atmospheric Administration.

Carss, D.N. (1990) Concentrations of wild and escaped fishes immediately adjacent to fish farm cages. *Aquaculture*, **90**, 29–40.

Christiansen, N.O. & Larsen, J.L. (1983) Microbiological and hygienic problems in marine aquaculture: experience and recommendations. In: *Diseases of Commercially Important Marine Fish and Shellfish* (ed. J.E. Stewart), pp. 49–53. ICES, Copenhagen.

Coyne, R., Hiney, M. O'Connor, B., Kerry, J., Cazabon, D. & Smith, P. (1994) Concentration and persistence of oxytetracycline in sediments under a marine salmon farm. *Aquaculture*, **123**, 31–42.

Cravedi, J.-P., Chouber, G. & Delous, G. (1987) Digestibility of chloramphenicol, oxolinic acid and oxytetracycline in rainbow trout and influence of these antibiotics on lipid digestibility. *Aquaculture*, **60**, 133–41.

Dahlback, B. & Gunnarsson, L.A.H. (1981) Sedimentation and sulfate reduction under a mussel culture. *Marine Biology*, **63**, 269–75.

Davies, J. & Smith, D.I. (1978) Plasmid-determined resistance to antimicrobial agents. *Annual Review of Microbiology*, **32**, 469–518.

Elmund, G.K., Morrison, S.M., Grant, D.W. & Nevins, M.P. (1971) Role of excreted chlortetracycline in modifying the decomposition process in feed lot wastes. *Bulletin of Environmental Contamination and Toxicology*, **6**, 129–32.

Ervik, A., Thorsen, B., Erikson, V., Lunestad, B.T. & Samuelsen, O.B. (1994) Impact of administering antibacterial agents on wild fish and blue mussels *Mytilus edulis* in the vicinity of fish farms. *Diseases of Aquatic Organisms*, **18**, 45–51.

FAO/WHO (1990) Evaluation of certain veterinary drug residues in food, 1990. *Thirty-sixth Report of the Joint FAO/WHO Expert Committee on Food Additives*. World Health Organization, Geneva.

Gowen, R.J., Bradbury, N.B. & Brown, J.R. (1985) The ecological impact of salmon farming in Scottish coastal waters: a preliminary appraisal. ICES C.M. 1985/F:35.

Grave, K., Engelstad, M., Søli, N.E. & Håstein, T. (1990) Utilization of antibacterial drugs in salmonid farming in Norway during 1980–1988. *Aquaculture*, **86**, 347–58.

Gray, J.P. & Herwig, R.P. (1994) Microbial response to antibacterial treatment in marine microcosms. In: *Environmental Fate and Effects of Aquacultural Antibacterials in Puget Sound* (eds D.P. Weston, D.G. Capone, R.P. Herwig & J.T. Staley). Appendix 5, Report from the University of California at Berkeley to the National Oceanic and Atmospheric Administration.

Gray, J.P., Weston, D.P. & Herwig, R.P. (1994) Antibacterial resistant bacteria in surficial sediments near salmon net-cage farms in Puget Sound, Washington. *Environmental Fate and Effects of Aquacultural Antibacterials in Puget Sound* (eds D.P. Weston, D.G. Capone, R.P. Herwig & J.T. Staley). Appendix 4, Report from the University of California at Berkeley to the National Oceanic and Atmospheric Administration.

Hall, P. & Holby, O. (1986) Environmental impact of a marine fish cage culture. ICES CM 1986/F:46.

Hansen, P.K., Lunestad, B.T. & Samuelsen, O.B. (1992) Effects of oxytetracycline, oxolinic acid, and flumequine on bacteria in an artificial marine fish farm sediment. *Canadian Journal of Microbiology*, **38**, 1307–12.

Hastings, T.S. & McKay, A. (1987) Resistance of *Aeromonas salmonicida* to oxolinic acid. *Aquaculture*, **60**, 133–41.

Holmer, M. & Kristensen, E. (1992) Impact of marine fish cage farming on metabolism and sulfate reduction of underlying sediments. *Marine Ecology Progress Series*, **80**, 191–201.

Hsu, C.-H., Hwang, S.-C. & Liu, J.-K. (1992) Succession of bacterial drug resistance as an indicator of antibiotic application in aquaculture. *Journal of the Fisheries Society of Taiwan*, **19**, 55–64.

Husevåg, B., Lunestad, B.T., Johannessen, P.J., Enger, Ø. & Samuelsen, O.B. (1991) Simultaneous occurrence of *Vibrio salmonicida* and antibiotic resistant bacteria in sediments at abandoned aquaculture sites. *Journal of Fish Diseases*, **14**, 631–40.

Jacobsen, P. & Berglind, L. (1988) Persistence of oxytetracycline in sediments from fish farms. *Aquaculture*, **70**, 365–70.

Kerry, J., Hiney, M., Coyne, R., Cazabon, D., NicGabhainn, S. & Smith, P. (1994) Frequency and distribution of resistance to oxytetracycline in microorganisms isolated from marine fish farm sediments following therapeutic use of oxytetracycline. *Aquaculture*, **123**, 43–54.

Kerry, J., Hiney, M., Coyne, R., NicGabhainn, S., Gilroy, D., Cazabon, D. & Smith, P. (1995) Fish feed as a source of oxytetracycline resistant bacteria in the sediments under fish farms. *Aquaculture*, **131**, 101–13.

Kitts, D.D., Yu, C.W.Y., Aoyama, R.G., Burt, H.M. & McErlane, K.M. (1992) Oxytetracycline degradation in thermally processed farmed salmon. *Journal of Agriculture and Food Chemistry*, **40**, 1977–81.

López-Jamar, E., Iglesias, J. & Otero, J.J. (1984) Contribution of infauna and mussel-raft epifauna to demersal fish diets. *Marine Ecology Progress Series*, **15**, 13–18.

Loyacano, H.A., Jr & Smith, D.C. (1976) Attraction of native fish to catfish culture cages in reservoirs. *Proceedings of the Annual Conference of the Southeastern Association of Game and Fisheries Commissioners*, **29**, 63–73.

Lunestad, B.T. & Goksoyr, J. (1990) Reduction in the antibacterial effect of oxytetracycline in sea water by complex formation with magnesium and chloride. *Diseases of Aquatic Organisms*, **9**, 67–72.

Meier, W., Schmitt, M. & Wahli, T. (1992) Resistance to antibiotics used in the treatment of freshwater fish during a 20 year period (1979–1988) in Switzerland. In: *Problems of Chemotherapy in Aquaculture: From Theory to Reality* (eds C.M. Michel & D.J. Alderman), pp. 141–50. Office International des Epizooties, Paris.

Meyer, F.P. & Schnick, R.A. (1988) A review of chemicals used for the control of fish diseases. *Reviews in Aquatic Science*, **1**, 693–710.

Michel, C. (1986) Practical value, potential dangers and methods of using antibacterial drugs in fish. *Revenue Scientifique et Technique, Office International des Epizooties*, **5**, 659–75.

Moats, W.A. (1988) Inactivation of antibiotics by heating in foods and other substrates: a review. *Journal of Food Protection*, **51**, 491–7.

Møster, G. (1986) *Bruk av Antibiotika i Fiskeoppdrett* (The Use of Antibiotics in Norwegian Fish Farming.) Sogn of Fjordane Distriktshøgskole, Sogndal.

NRC (1992) *Marine Aquaculture: Opportunities for Growth*. National Academy Press, Washington.

Neuvonen, P.J. (1976) Interactions with the absorption of tetracyclines. *Drugs*, **11**, 45–54.

Nygaard, K., Lunestad, B.T., Hektoen, H., Berge, J.A. & Hormazabal, V. (1992) Resistance to oxytetracycline, oxolinic acid and furazolidone in bacteria from marine sediments. *Aquaculture*, **104**, 31–6.

Oka, H., Ikai, Y., Kawamura, N., Yamada, M., Harada, K., Ito, S. & Suzuki, M. (1989) Photodecomposition products of tetracycline in aqueous solution. *Journal of Agriculture and Food Chemistry*, **37**, 226–31.

Okamoto, A. (1992) Restrictions on the use of drugs in aquaculture in Japan. In: *Problems of Chemotherapy in Aquaculture: From Theory to Reality* (eds C.M. Michel & D.J. Alderman), pp. 109–14. Office International des Epizooties, Paris.

Primavera, J.H., Lavilla-Pitogo, C.R., Ladja, J.M. & Peña (1993) A survey of chemical and biological products used in intensive prawn farms in the Philippines. *Marine Pollution Bulletin*, **26**, 35–40.

Ram, N., Ulitzur, S. & Avnimelech, Y. (1981) Microbial and chemical changes occurring at the mud–water interface in an experimental fish pond. *Bamidgeh*, **33**, 71–86.

Ram, N.M., Zur, O. & Avnimelech, Y. (1982) Microbial changes occurring at the sediment–water interface in an intensively stocked and fed fish pond. *Aquaculture*, **27**, 63–72.

Richards, R.H., Inglis, V., Frerichs, G.N. & Millar, S.D. (1992) Variation in antibiotic resistance patterns of *Aeromonas salmonicida* isolated from Atlantic salmon *Salmo salar* L. in Scotland. In: *Problems of Chemotherapy in Aquaculture: From Theory to Reality* (eds C.M. Michel & D.J. Alderman), pp. 151–8. Office International des Epizooties, Paris.

Robins-Brown, R.M., Gaspar, M.N., Ward, J.I., Wachsmith, I.K., Koornhof, H.J., Jacobs, M.R. & Thornsberry, C. (1979) Resistance mechanisms of multiple resistance pneumococci: antibiotic degradation studies. *Antimicrobial Agents and Chemotherapy*, **15**, 470–4.

Rogstad, A., Hormazabal, V., Ellingsen, O.F. & Rasmussen, K.E. (1991) Pharmokinetic study of oxytetracycline in fish. I. Absorption, distribution and accumulation in rainbow trout in freshwater. *Aquaculture*, **96**, 219–26.

Romero, P., González-Gurriarán, E. & Penas, E. (1982) Influence of mussel rafts on spatial and seasonal abundance of crabs in the Ria de Arousa, Northwest Spain. *Marine Biology*, **72**, 201–10.

Rosenthal, H., Weston, D., Gowen, R. & Black, E. (1988) Report of the *ad hoc* study group on the environmental impact of mariculture. *ICES Cooperative Research Report*, **154**.

Saitanu, K., Amornsin, A., Kondo, F. & Tsai, C.-E. (1994) Antibiotic residues in tiger shrimp (*Penaeus monodon*). *Asian Fisheries Science*, **7**, 47–52.

Samuelsen, O.B. (1989) Degradation of oxytetracycline in seawater at two different temperatures and light intensities, and the persistence of oxytetracycline in the sediment from a fish farm. *Aquaculture*, **83**, 7–16.

Samuelsen, O.B. (1992) The fate of antibiotics/chemotherapeutics in marine aquaculture sediments. *Problems of Chemotherapy in Aquaculture: From Theory to Reality* (eds C.M. Michel & D.J. Alderman), pp. 87–95. Office International des Epizooties, Paris.

Samuelsen, O., Torsvik, V. Hansen, P.K., Pittman, K. and Ervik, A. (1988) Organic waste and antibiotics from aquaculture. ICES CM 1988/F:14.

Samuelsen, O.B., Solheim, E. & Lunestad, B.T. (1991) Fate and microbiological effects of furazolidone in a marine aquaculture sediment. *The Science of the Total Environment*, **108**, 275–83.

Samuelsen, O.B., Torsvik, V. & Ervik, A. (1992a) Long-range changes in oxytetracycline concentration and bacterial resistance towards oxytetracycline in a fish farm sediment after medication. *The Science of the Total Environment*, **114**, 25–36.

Samuelsen, O.B., Lunestad, B.T., Husevåg, B., Hølleland & Ervik, A. (1992b) Residues of oxolinic acid in wild fauna following medication in fish farms. *Diseases of Aquatic Organisms*, **12**, 111–19.

Samuelsen, O.B., Lunstad, B.T., Ervik, A. & Fjeldes, S. (1994) Stability of antibacterial agents in an artificial marine aquaculture sediment studied under laboratory conditions. *Aquaculture*, **126**, 283–90.

Schnick, R.A. (1991) Chemicals for worldwide aquaculture. In: *Fish Health Management in Asia-Pacific. Report on a Regional Study and Workshop on Fish Disease and Fish Health Management*, pp. 441–67. *ADB Agriculture Department Report Series*, **1**. Network of Aquaculture Centres in Asia-Pacific, Bangkok.

Smith, P., Donlon, J., Coyne, R. & Cazabon, D.J. (1994a) Fate of oxytetracycline in a fresh water fish farm: influence of effluent treatment systems. *Aquaculture*, **120**, 319–25.

Smith, P., Hiney, M. & Samuelsen, O. (1994b) Bacterial resistance to antimicrobial agents used in fish farming; a critical evaluation of method and meaning. *Annual Review of Fish Diseases*, **4**, 273–313.

Stickney, R.R. (1988) Aquaculture on trial. *World Aquaculture*, **19**, 16–18.

Stinson, S.C. (1987) Animal health products recovering momentum. *Chemical and Engineering News*, **65**, 51–68.

Thorpe, J.E., Talbot, C., Miles, M.S., Rawlings, C. & Keay, D.S. (1990) Food consumption in 24 hours by Atlantic salmon (*Salmo salar* L.) in a sea cage. *Aquaculture*, **90**, 41–7.

Timoney, J.F. & Port, J.G. (1982) Heavy metal and antibiotic resistance in Bacillus and Vibrio from sediments of New York Bight. In: *Ecological Stress and the New York Bight: Science and Management* (ed. G.F. Mayer), pp. 235– 48. Estuarine Research Federation, Columbia, South Carolina.

Torsvik, V.L., Sørheim, R. & Goksøyr, J. (1988) Antibiotic resistance of bacteria from fish farm sediments. ICES CM 1988/F:10.

Tunison, A. & McCay, C. (1937) The nutrition of trout. *Cortland Hatchery Report*, **6**, New York State Department of Conservation, Albany.

VKI (1976) Forskellige driftsparameters indflytelse på forureningen fra dambrug (The influence of different operational parameters on the pollution from fish farms). Rapport til Teknologiradet., Vandkvalitetsinst & Jydsk Teknologisk. Inst., Horsholm, Denmark.

Weston, D.P. (1990) Quantitative examination of macrobenthic community changes along an organic enrichment gradient. *Marine Ecology Progress Series*, **61**, 233–44.

Weston, D.P. (1994) Patterns of antibacterial use in salmonid net-cage culture in Puget Sound. In: *The Environmental Fate and Effects of Aquacultural Antibacterials in Puget Sound* (eds D.P. Weston, D.G. Capone, R.P. Herwig & J.T. Staley). Appendix 1. Report from the University of California at Berkeley to the National Oceanic and Atmospheric Administration.

Weston, D.P., Capone, D.G., Herwig, R.P. & Staley, J.T. (1994) *The Environmental Fate and Effects of Aquacultural Antibacterials in Puget Sound*. Report from the University of California at Berkeley to the National Oceanic and Atmospheric Administration.

Wood, S., McCashion, R. & Lynch, W. (1986) Multiple low-level resistance in *Aeromonas salmonicida*. *Antimicrobial Agents and Chemotherapy*, **23**, 475–83.

Yndestad, M. (1992) Public health aspects of residues in animal products: fundamental considerations. In: *Problems of Chemotherapy in Aquaculture: From Theory to Reality* (eds C.M. Michel & D.J. Alderman), pp. 494–510. Office International des Epizooties, Paris.

Chapter 7
Reductions in Wastes from Aquaculture

Simon J. Cripps   *Rogaland Research, Stavanger, Norway*
Liam A. Kelly   *Heriot Watt University, Edinburgh, UK*

7.1 INTRODUCTION

Statistics reveal that throughout most of the world, the aquaculture industry is expanding (FAO 1992), in terms of both the overall production and intensity of aquaculture operations. New technology, such as offshore cages, and improved production techniques, are allowing the industry to expand into previously undeveloped sites. Additionally, space limitations or environmental requirements have forced some producers into new sites in marginal locations that may be expected to have good quality water. The potential pollutant mass flow and concentration in the waste water will normally increase as production increases in both volume and intensity. The impact of waste water will be particularly pronounced in areas with clean waters, or where the carrying capacity of the site is approached. At this point, control of wastes is required.

The potential impacts of the aquaculture industry have been described by many authors including Pillay (1992), and include those resulting from feed derived wastes (Gowen 1991), genetic pollution (Crozier 1993), introduction of exotic species (GESAMP 1991), predator conflicts (Howell & Munford 1991), chemical addition (Beveridge *et al.* 1994), coastline degradation (Pillay 1992) and social conflicts (Pollnac 1992). Careful site selection and good management can reduce many of these impacts, but by definition cage systems are open, and therefore possess a dynamic rather than static relationship with the surrounding environment. Flow-through tank systems, though partially isolated and potentially easier to adapt to the environmental requirements of the cultured organism in a particular location, are also both dependent on, and have an impact upon, the surrounding environment.

Pollution by nutrients or pathogens in waste water is currently perceived, rightly or wrongly, to be one of the main sources of impact on the environment from aquaculture facilities. Waste from a farm can degrade the environment and conflict with other aquatic resource users, or the waste outputs may be incorrectly suspected of having a greater effect than actually occurs. The consequences are, however, the same; the development of an aquaculture site now demands the use of appropriate planning and adequate waste reduction technology. Whilst existing sites may, for regulatory purposes, require 'retrofitted' (i.e. added to the site after the construction period) pollution abatement equipment in response to changing requirements, either by external legislation and/or local pressures on the use of the aquatic resource. It is clear that, whatever the reason, the demand for methods of reducing the wastes in effluents from aquaculture facilities is increasing.

This review aims to define the problems associated with treating waste water and describe procedures for the collection of waste. Relevant, representative treatment methods and their theoretical bases will be discussed. Alternative waste reduction methods and possible future trends in treatment will also be considered. It is hoped that a better understanding of the waste problems, in terms of both quality and quantity, will assist both site operators and local planning decisions regarding aquaculture developments.

7.2   TREATABLE WASTE WATER CHARACTERISTICS

The study and treatment of domestic waste water is a relatively mature field, having been conducted intensively for well over a century (Tchobanoglous & Burton 1991). Conversely, aquaculture effluent characterisation and treatment is a relatively new field. The properties of this unusual type of industrial effluent are not conducive to simple treatment.

There are two main interlinked problems associated with the treatment of waste water from aquaculture facilities, low concentrations of potential pollutants and high flow rates (Cripps 1994). Liquid effluents with a higher contaminant concentration are generally easier to treat, in terms of producing a reduction in pollutant concentration. In the water treatment field it is therefore common to describe effluent content in concentration units, for example mg/l, rather than loadings, such as the weight of nutrient loss to the recipient per kg of fish cultured, or per kg of feed used.

Typical aquaculture pollutant concentrations, in comparison with other common effluents, are low (Fig 7.1). The concentration of contaminants in aquaculture effluents, combined with high flow rates, thus presents treatment problems. The total mass flow emanating from a farm may, therefore, in time, become high. Cho *et al.* (1991) state that the traditional strategy of 'dilution is the solution' may be environmentally misleading. Perhaps of greater influence to the recipient water body is the total loading during a given time period, e.g. kg of waste substance per day.

Two differing considerations therefore exist regarding the design and operation of waste treatment systems. Reduction in effluent concentration is considered important within the treatment field, whilst reduction in effluent loading rates is important within the environmental field. These two viewpoints do not necessarily have similar solutions and outcomes. The effect of waste treatment, loadings and effluent concentration will be discussed in this review.

Four main components of the waste water are of environmental and treatment

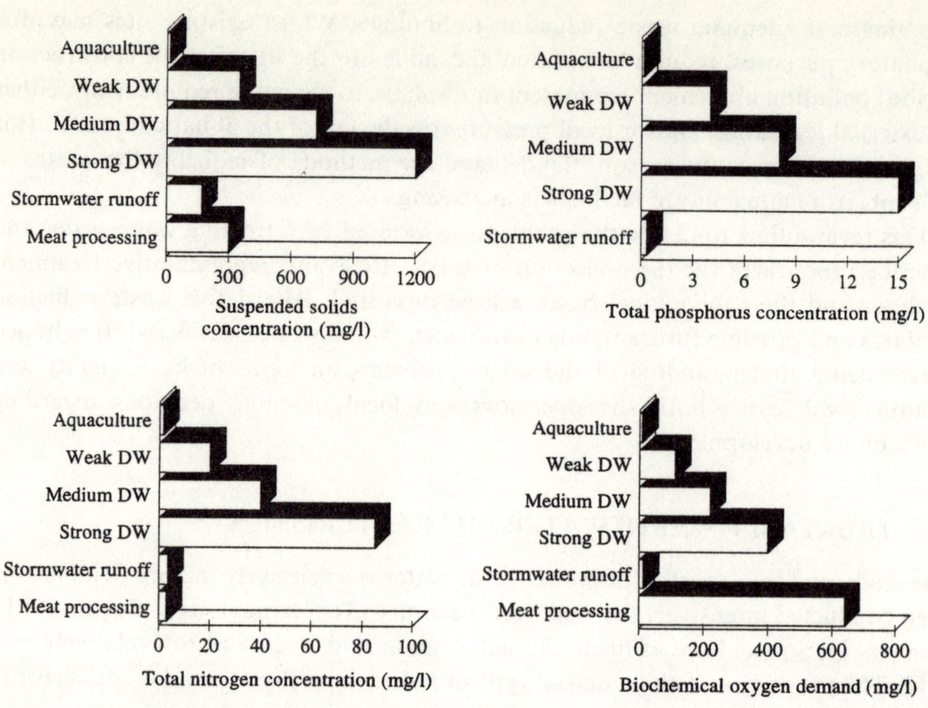

Fig. 7.1   Aquaculture waste water contaminant concentration in comparison with other industrial waste waters. Aquaculture (Cripps 1994); DW (domestic waste water) (Tchobanoglous & Burton 1991); storm water runoff and meat processing (Henry & Heinke 1989).

interest; nutrients, biochemical oxygen demand (BOD), suspended solids and pathogens. Other constituents, such as CO_2, may also be of interest in particular locations, or in specialised forms of culture, such as recycle systems.

7.2.1   Nutrients

Nitrogen (N) in sea water, and phosphorus (P) in fresh water, are usually the nutrient compounds which primarily limit growth in natural ecosystems. For similar reasons, they are also important feed components of culture organisms, required for normal growth (Shepherd & Bromage 1988). The source of the bulk of N and P found in many intensive systems is therefore feed. To ensure maximum growth, these nutrients are normally incorporated into the feed in excess of normal requirements (Wiesmann *et al.* 1988). Latterly, however, the tendency has been to produce feeds with N and P contents more closely aligned to the dietary needs of the cultured species, in order to reduce wastage. In addition to unused nutrients resulting from food which is not eaten, either regurgitated or unmetabolised, this excess results in nutrient discharges into the environment (NCC 1990). It is these nutrients that can cause the greatest environmental impacts, such as eutrophication (Pillay 1992; Welch & Lindell 1992).

The problem associated with the treatment of these nutrients, as previously stated, is that they typically occur in low concentrations in effluents, although their mass flow to the recipient water body may be substantial. Table 7.1 indicates the range of four contaminant concentrations in aquaculture waste water. These values, when compared with other industrial wastes, can be seen to be low (Fig. 7.1). On first inspection they may be considered too low to warrant mitigation. Large proportional reductions in these nutrients by water treatment technology would, as a result of their already low concentrations, produce only small changes in absolute concentration. Analysis of mass flow or loading data, however, indicates that treatment is indeed often required.

Table 7.1 Reported salmonid farm effluent concentrations (Cripps 1994).

Facility	SS (mg/l)	TP (mg/l)	TN (mg/l)	BOD (mg/l)	Source
21 EIFAC farms	9	—	—	5	Alabaster (1982)
Typical in Norway	3	0.100	0.5	—	Bergheim et al. (1991b)
Rogaland, Norway	1.6–14.1	0.080–0.270	0.43–0.70	—	Bergheim et al. (1993)
Northern Sweden	[6.9]	0.110	0.70	—	Cripps (1995)
N. Ireland		0.110[a]	0.531[a]	—	Foy & Rosell (1991a)
Typical	5–50	0.050–0.260	0.5–5.0	5–20	Muir (1982)
Finland	—	0.055	—	—	Mäkinen et al. (1988)
> 31 UK farms	11.1	0.082		4.00	Solbé (1982)
Typical in Denmark	5–50	0.050–0.150	0.5–4.0	3–20	Warrer-Hansen (1982)
Typical values	14	0.125	1.4	8	To Fig. 1

[ ] = unpublished data
[a] = increase from farm inlet to outlet

The N and P loads of intensive salmonid aquaculture systems have been described by several workers and compiled in a comprehensive study of the interactions between aquaculture and the environment (NCC 1990, Fig. 7.2). These loadings represent only a small proportion of the total loading to the recipient from other activities such as agricultural runoff and sewage waste. Ackefors & Enell (1990), for example, estimated that 0.6% and 0.2% of the Swedish P and N loads, respectively, originated from aquaculture activities. Kronvang et al. (1993) indicated in a nation-wide study of freshwater resources that in Denmark, traditionally viewed as an area heavily used for aquaculture, the N and P discharged from fish farms to freshwater recipients was respectively 2% and 4% of total loadings.

Whilst concentrations and loadings may be considered small in comparison with other sources, they nevertheless represent a significant input. The total input from Norwegian farms was estimated at 2800 t of P and 13 700 t of N during 1989 (Braaten 1991) and more conservatively, from the six Nordic countries, as 2400 t of P and 14 200 t of N during the same year (Enell & Ackefors 1991). As awareness of envir-

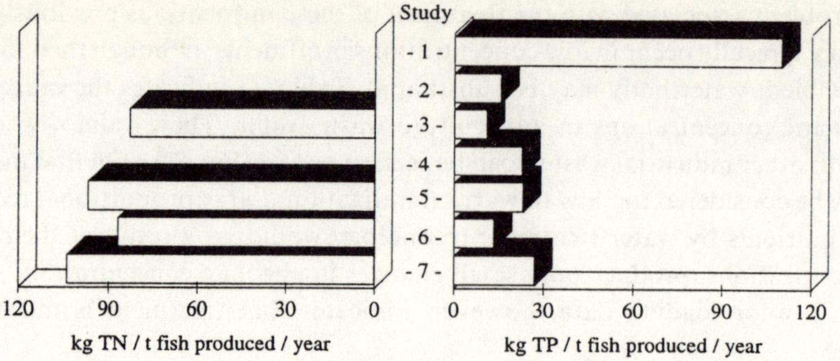

Fig. 7.2 Total nitrogen (TN) and total phosphorus (TP) loadings (upper values of ranges are shown) reported in various studies and compiled by NCC (1990). Study 1, mainly European salmonids (Alabaster 1982); study 2, UK salmonids (Solbé 1982); study 3, rainbow trout, dry feed, ponds and tanks (Warrer-Hansen 1982); study 4, rainbow trout, dry feeds (Ketola 1982); study 5, Polish rainbow trout, moist feed, cages (Penczak *et al.* 1982); study 6, Swedish rainbow trout, dry feed, cages (Enell & Lof 1983); study 7, Scottish rainbow trout, dry feed, cages (Phillips 1985).

onmental priorities increases, the aquaculture industry must expect to keep pace with other industries by reducing both inputs and outputs correspondingly.

The aquaculture industry also has other less altruistic reasons for reducing waste nutrients. The impacts which they cause may have profound localised effects, both on the environment and on the farm itself, and in both environmental and financial terms. The impacts of these outputs are considered elsewhere in this volume.

It has been shown (Ackefors & Enell 1994) that a large proportion of the nutrients is not used for the purpose for which they were added, i.e. the growth of the culture organism (Fig. 7.3). This is an indication of inefficiency. There is, therefore, scope for the application of prophylactic technology, e.g. feed and feeding optimisation, which should not only improve farm economies but also reduce environmental impacts. The need to combine treatment technology and management decisions to reduce waste nutrient losses is therefore desirable in all intensive aquaculture operations.

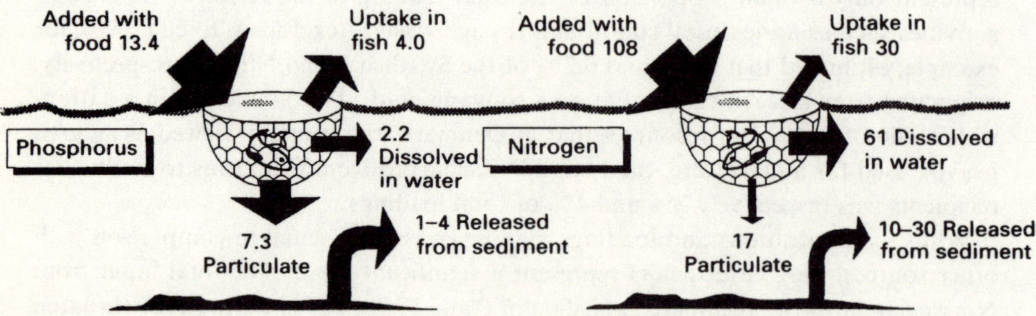

Fig. 7.3 Phosphorus and nitrogen loadings (kg/t fish produced per season) from a cage fish farm (from Ackefors & Enell 1994).

7.2.2   BOD, dissolved oxygen and biodegradable organic matter

Waste organic matter, in addition to that occurring naturally in the recipient water body, will be used by microbes as a source of energy. The breakdown or oxidation of these organic materials will occur both within the culture system and downstream of the outfall, and requires oxygen. Though the chemical pathways are complex and the quantity of oxygen required must be estimated empirically, it is proportional to the organic concentration. The amount of organic matter in an effluent may be measured directly as total organic carbon (TOC), but this value does not indicate the proportion of the organics which is biodegradable (Henry & Heinke 1989). The quantity of biodegradable organics present is then estimated indirectly by measuring the quantity of oxygen used by microbes, the BOD, to convert organic matter to the end products of CO_2 and water in a closed system. This method has two advantages, in that the organic content of the effluent is determined for treatment purposes, and the reduction in the ambient oxygen concentration is estimated for environmental impact assessments.

The measurement of BOD, usually as BOD_5 or BOD_7, as described by Tchobanoglous & Burton (1991), has several limitations, particularly in determinations of saline waste water quality. There is a lack of information relating to chemical oxygen demand (COD), which may be a more significant measure of polluting potential in many situations. Landau (1992) noted that, in addition to the microbial breakdown of organics, an aquaculture facility will also produce a BOD resulting from the depletion of oxygen, as a result of the metabolic requirements of high density culture organisms and from endemic aquatic plants respiring in low light conditions.

7.2.3   Suspended solids and particles: particle size and composition

Suspended solids (SS) concentration is an important and basic measure of effluent quality, widely adopted in waste water engineering. SS in aquaculture effluents are derived from a variety of sources, being carried into the farm in the inlet and produced within the farm, mainly as uneaten or regurgitated food and as faeces. Maximum SS concentrations at the outlet of temperate aquaculture facilities are typically about 14 mg/l, though this value fluctuates considerably, as described in [7.2.4] (Table 7.1).

A number of physical, mechanical and chemical problems result from high or prolonged SS loads being released into the environment, or retained within the farm. The physical problems experienced downstream include smothering of the benthos, water opacity and clogging of the respiratory apparatus of fish. Chemical problems arise from the composition of the solids. The majority of P exported from fish farms is bound to the particulate fraction (Ackefors & Enell 1994). This fraction may sediment out and, under certain conditions, may be later released into the environment (Kelly 1992). It is, though, the dissolved fraction which may have more immediate effects on water quality (NCC 1990; Kelly 1993). Conversely, less N may be in the particulate form than dissolved (Ackefors & Enell 1994), with the proportion of total N in the particulate form ranging at least between 20% (Cripps 1995) and 32% (Bergheim *et al.* 1993). Reasons for the variation will be discussed in [7.2.4]. Much of the biodegradable organic matter, which produces a BOD and reduces dissolved oxygen (DO) concentrations, is also present in the particulate fraction.

Methods employed to treat aquaculture wastes with low pollutant concentrations, high flow rates and particle bound pollutants, therefore, usually function by the separation of particles from the carrying flow. The particle number concentration in aquaculture wastes appears high. Cripps (1993) measured 1833–1 880 000 particles/l within the size range 8.4–269.2 µm. The majority of these particles, however, were small (< 30 µm diameter), an unquantified proportion of which were probably carried into the farm in the inlet. Such small particles, present at low concentrations, are difficult to remove from the effluent stream. Peaks associated with factors such as feed constitution were not evident. Smaller particles were simply present in larger numbers (Fig. 7.4). The transport of pollutants is size-specific: single larger particles within a defined size range have a considerably greater volume, and therefore potential to transport pollutants, than several hundred smaller particles (Fig. 7.4). So, whilst technical limitations on treatment methods result in the removal of a relatively small number of larger particles, it is these particles which have the greater pollutant-carrying potential, and to which most treatment effort should be directed.

Treatment efficiency, in terms of nutrient reduction, has been found to be proportional to separation effort, i.e. no specific particle size fraction was more important to remove, and therefore the smaller the pore size used on a mechanical

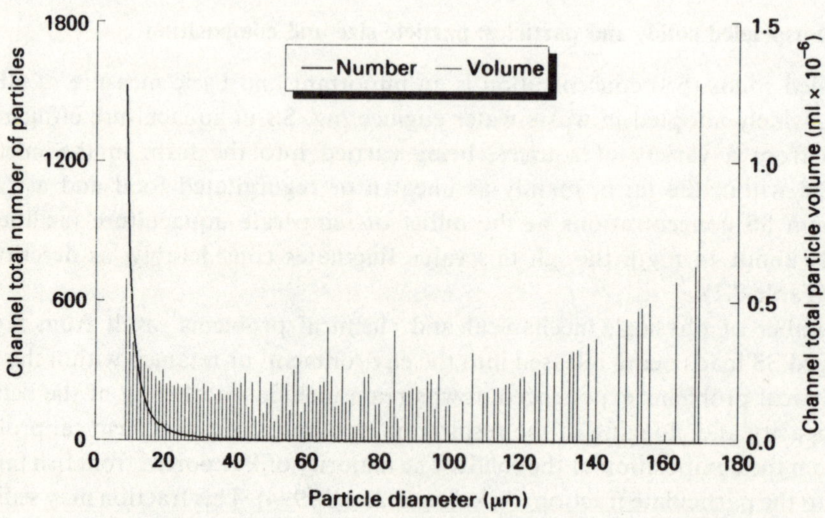

Fig. 7.4 Size distribution of suspended particle numbers and volumes in the effluent from a salmon hatchery (from Cripps 1993).

separator screen, the greater will be the pollutant reduction (Cripps 1995). This is a conclusion which is not self-evident in several other industrial waste treatment processes.

7.2.4 Component variability

The specific characteristics of an effluent vary considerably, both with time and between different farms, so aquaculture effluents cannot usually be precisely quantified. Nutrients such as N and P are partially associated with the suspended particles in the effluent, though the degree of association varies considerably (Table 7.2). Various factors may affect the association. Foy & Rosell (1991b) found that the proportion of nutrients in the particulate form increased with temperature, whilst Kelly *et al.* (1994) found that overall outputs of waste also increase with water temperature. Feeding rate of the culture species and feed composition will also have an effect.

Table 7.2 Variation in the percentage of nutrients associated with the particulate fraction.

Percentage nutrient concentration in the particulate fraction		Particle size (µm)	Reference
TN	TP		
26.6 (SIN)	30	> 0.45	Foy & Rosell (1991a)
7–32	47–84	> 60	Bergheim *et al.* (1993)
20	25	> 0.45	Cripps (1995)

The residence time of the particles within the culture system and mechanical agitation or damage to the particles will undoubtedly also influence nutrient partitioning, although few studies have considered this problem. Nutrients within particles which remain in suspension or settle within the culture system will leach into solution. Irrespective of chemical properties such as pH, temperature, ionic composition and nutrient concentration already in solution, several physical/mechanical factors are known to increase the quantity of nutrients moving from the particulate to the dissolved fraction (Pontius 1990). These include the increased flow of water around particles, causing scouring and increasing the nutrient concentration gradient; increased agitation, leading to break-down of particles to smaller sizes and consequently producing particles with slower settling velocities (Boersen & Westers 1986); and, as a function of this process, an increase in the surface area: volume ratio.

These factors will vary considerably between different farms, owing to differences in farm design, operation and management. They combine to produce a wide range of effluent characteristics, which vary with both location and time. Design criteria which have an impact upon waste characteristics include choice of tank design, the distance between the culture vessels and the treatment devices, flow rates, building materials (e.g. concrete has a higher surface roughness than plastic (Ekbäck 1975)), pump positioning and depth under cages. Operational and management criteria which may also determine waste characteristics are primarily influenced by choice of feed type and the protocol for delivery to the stock, but also include local decision-making, such as the frequency of cleaning and flushing of production units.

7.2.5 Effluent control strategy

Several strategies exist for the reduction of wastes in aquaculture systems. The choice of strategy must, however, be customised to the operation of the particular culture system. Culture technique and the species cultured have a considerable influence on the timing and nature of effluent outputs (Table 7.3). A consequence of this is that almost all waste treatment devices will require a large element of local planning. This may go beyond engineering considerations, and may in some instances, such as where a device is retro-fitted, require revision of existing infrastructure and existing work practices, in order to obtain the best possible configuration of waste output and treatment. The examples chosen below present a series of different hydraulic and waste dynamic conditions, and consequent management problems. The clear implication is that the question of what constitutes a waste treatment problem has many different meanings in different aquaculture systems (Table 7.3).

Quantification of the waste 'signal' from the facility is therefore an important requirement. Data regarding other factors, such as size of the facility and waste

Table 7.3   Defining the waste problems for land-based acquaculture systems.

Species	System	Culture mode	Water flow	Waste output type
Salmon (SW & FW) (*Salmo Salar*)	Tanks	Batch	Continuous	Daily pulsed peak
Salmon (SW & FW) (*Salmo salar*)	Self cleaning tanks	Batch	Continuous	Continuous dilute
Trout (FW) (*Oncorhynchus mykiss*)	Ponds/ raceways	Continuous	Continuous	Continuous dilute
Shrimp (SW) (*Penaeus monodon*)	Ponds	Batch	Periodic	Occasional pulsed/Harvest peak
Carp (FW)	Ponds	Batch	Still water	Harvest peak
Catfish (FW) (*Clarias gariepinus*)	Tanks	Continuous	Periodic	Harvest peak

loadings, are of little use if treatment effort does not match periods of highest waste release. An example of this problem (Boersen & Westers 1986) may be seen in juvenile salmonid farms, where peaks in wastes, accompanied by rapid changes in water discharge, occur during short periods of the day when tanks are flushed (Fig. 7.5). This problem is not dissimilar to that of urban sewage systems, which experience sharp changes in the quality and quantity of waste to be treated during a working day. The more complex urban treatment systems, however, have greater flexibility of operation as a result of closed, defined planning.

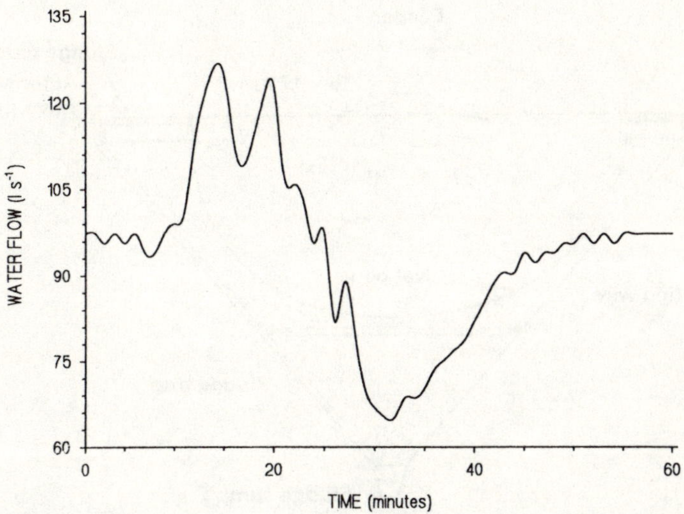

Fig. 7.5   Variation in water discharge during tank cleaning at a Scottish smolt production site (L.A. Kelly, unpublished data).

7.3   WASTE COLLECTION TECHNOLOGY

The low overall concentration of potential pollutants in the waste water from aquaculture facilities also hampers efficient collection prior to treatment. Land and sea-based systems have, in addition to this, further problems, due to their differing physical configuration.

7.3.1   Cage systems

In cage farms, a flow is maintained through the cages, primarily to ensure a surplus of DO and to carry away wastes. Removal of wastes by dispersion through the cage sides and bottom effectively produces a diffuse source of pollution, and so wastes are difficult to collect prior to treatment or disposal. Fluctuations in the direction and strength of water movement through open systems are an aid to the wider dispersion of wastes if no treatment is required, but also further hamper collection. The need to distribute fish and food evenly throughout the cage also increases dispersal. Waste collection or removal from cages is notoriously difficult, and so only a few methods have been tested, with varying success. These include collectors, closed cages, water column and sediment pumps. Bergheim *et al.* (1991a) described a cage sludge collection system with a horizontal area of 50 m^2, corresponding to 40% of the cage area (Fig. 7.6). It was reported to have collected 75–80% of the settleable particles during a 4 month period. The collected material, with a dry matter content of 5–15%, was pumped daily from a sump at the bottom of the trap.

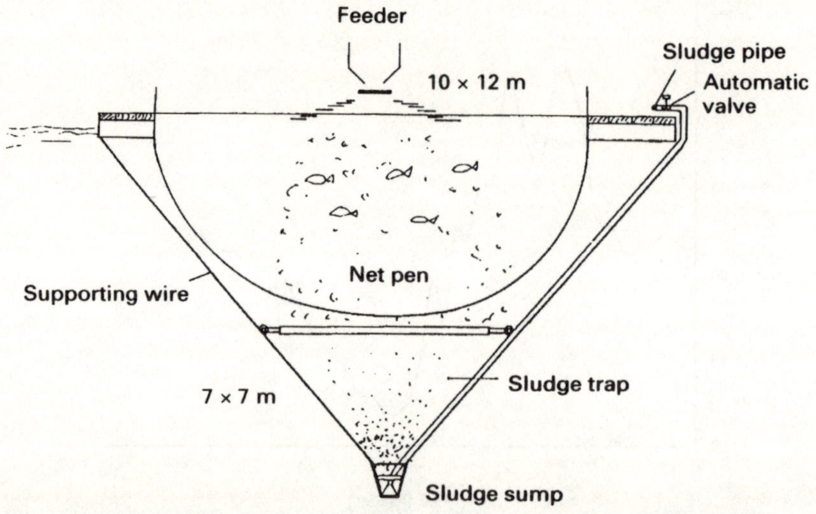

Fig. 7.6   Cross section of a cage sludge collection system (from Bergheim *et al.* 1991a).

The *Lift Up* system (Lift Up A/S, N-5640 Eikelandsosen) for the collection and subsequent removal of waste particles and dead fish has been described as operating efficiently (Braaten 1991, 1992). The system was designed to meet Norwegian Government regulations regarding the release of antibiotics in waste food into the environment. The system comprises a coarse mesh net around the outside of the cage. The lower part of the net forms a finer mesh net funnel. Waste material is pumped intermittently to a coarse screen located above the water level on the cage collar. Independent test results estimated that almost 100% of waste pellets larger than 6 mm were collected within minutes (Braaten 1992). The loss of antibiotics was reduced by 80%.

Whilst such collection systems do appear to be relatively successful, their efficiency under rough weather conditions may be reduced and thus their application restricted to periods of good weather and locations sheltered from strong water currents. Water flow through a mesh finer than that of the cage will be relatively restricted, and care must be taken to prevent a decline in water quality within the cage. Prototype closed cage systems have been developed, though there is little published information describing their effectiveness. They are much in demand by cage farmers, but their use has to date been limited. Problems of cost, reliability, and ease of operation may still need to be overcome.

Water flowing into cages may occasionally be of poor quality and the wastes flowing out considered to pollute the environment. Closed cages in which the containing structure is made of a solid material, such as high density polyethylene rather than netting, would appear to solve these problems. The inlet water can be drawn by pump from a location where the water is of a good quality and a suitable temperature, e.g. warmer bottom water in a lake. The water level in the cage is only a few centimetres above the outside water level, so the height to which the water must be pumped is negligible, irrespective of the depth of the water from which it is pumped. There will, though, be friction losses in the pipes. All of the outlet water can then be directed to a treatment facility prior to discharge, so the system may be considered to be closed. Another advantage is that periodic disease treatments of fish can be achieved with more accurate estimates of dosages administered.

The technical problems compared with traditional net pen designs, however, remain considerable; sufficient water exchange must be achieved at all times using pumps. The water must be distributed evenly throughout the cage and waste material must be carried away efficiently. The weight of the small head of water within the cage required to maintain the shape of the bag must be supported by the floating collar. The solid sides of the cage will present a greater barrier to external currents and wave forces than mesh cages, so the design needs to be sturdy and strongly anchored. During the past 5 years, economic constraints have limited the development of this system, but at present industrial and research interest is increasing again. At inland sites, particularly in slower-flowing lakes, this system would appear to be particularly appropriate.

More speculative still, but requiring further study, is the use of pumps below cages to remove particles. Theoretical modelling studies of flow dynamics around cage units indicate that a large, but as yet unquantified, proportion of settleable particles can be collected, using the correct combination of pump speed and inlet location. Resulting particle contour maps indicate that the particles can be selectively drawn towards the pump inlet.

Sediment pumps may also be a means of removing settled particles under cages, especially in heavily polluted regions. The frequency of renovation required will depend on the farm size and accumulation of organics (Gowen *et al.* 1991).

7.3.2 Tank systems

Flow rates in tank-based systems are also established primarily to maintain a high concentration of DO. This produces waste water in which the contaminant concentration is low, but the total mass flow loading may be high. Land-based systems are particularly suitable for the application of various methods of waste water treatment, because it is relatively easy to collect and direct an effluent flow emanating from a tank outlet pipe, compared with cage systems. The ability to treat farm waste waters to reduce the environmental impact is a considerable advantage of land-based facilities. There are several methods available for the concentration, collection and removal of solid wastes produced in tanks prior to treatment. These include flushing, self-cleaning, sediment traps, particle concentrators and *in situ* treatment devices.

The traditional method of waste collection in tanks, that of intermittent tank cleaning, is probably still the most common. Waste that builds up during a period of hours, days or weeks, depending on the hydrodynamics and roughness of the construction materials, is scrubbed from the tank floor and walls. This matter is resuspended so that it can be carried out of the tank in the outlet stream. To speed the removal, the tank may be flushed by increasing the inflow rate and lowering the water level. This method is not conducive to a good culture environment, but may confer treatment advantages.

The need to remove settled matter mechanically from a tank is an indication that the design of the tank (Wheaton 1977), or its hydrodynamic management (Cripps & Poxton 1992), is not optimal. For matter to settle, quiescent zones are required. Quiescent zones in a tank are an indication that there are dead regions in which the water is not mixed as frequently as in other parts of the tank (Burrows & Chenoweth 1955; Cripps & Poxton 1993). This reduces water quality in these areas (Rosenthal *et al*. 1982) and increases the short-circuiting of water (Cholette & Cloutier 1959) directly from the inlet to the outlet without being fully used. Only dead areas required deliberately to trap particles should be present in a tank, and should preferably be partitioned from the culture species. Resuspension of suspended matter can reduce the DO concentration and cause gills to become clogged, particularly in high density cultures, resulting in stress or death. The practice of scrubbing tanks is also stressful in itself.

From a treatment aspect, the intermittent production of a peak, or plug, in the suspended solids concentration of the effluent can be beneficial. Treatment equipment, such as particle screens, can be operated in a low-use mode for the majority of the time. Back-washing of the microscreens can be reduced, because there are fewer particles in the effluent flow. A correspondingly smaller volume of a more concentrated sludge can be produced. When the tanks are cleaned, the treatment effort can be increased for the period corresponding to the increased particle load. In the case of rotating micro-screens for example, an increase in the rotation speed and the backwashing volume would be required. The sludge production quantity using this strategy would be variable, but could be buffered using storage tanks prior to dewatering techniques, such as sedimentation, which function best at a constant, low flow rate.

Similar advantages can be achieved using particle traps. These are areas of a tank in which the local residence time has been increased. Flow in these areas is decreased, allowing particles to settle out. The most common types of particle traps employ in-tank structures to deflect the main flow and produce quiescent areas.

There are two main strategies used to intermittently empty these traps and send the material for treatment. Firstly, the sediment in the traps can be flushed using a large flow, or even mechanically scrubbed or scraped (Haynes 1975). The waste that flows out with the main outlet water flow has to be separated again using downstream treatment devices. Secondly the waste can be vented through a separate outlet, designed to carry only the sludge and not the main flow (Ulgenes & Eikebrokk 1992). The waste is then sent to a treatment unit which is separate, or temporarily dis-connected, from the main effluent flow. The second method is potentially more efficient, because the relatively concentrated particles, compared with the main effluent, are retained separate, and therefore the quantity of waste water requiring intensive treatment is small. The former method, in which the particles are resus-pended, produces a peak in SS relative to the average SS concentration, but the flow of water requiring treatment is large.

Several examples of sediment traps are available in the literature. Westers (1991) describes how a knowledge of flow dynamics and settling velocities can be used to design an efficiently functioning trap, though the resulting culture conditions may not have been suitable for all fish. Several baffles were located across the raceways. A gap between the baffles and the tank floor created regions of high flow, which carried the particles along the bottom towards and into a fish-free collection zone (Fig. 7.7). Particle interception was estimated at 70–80%, but the particle size range captured was not quantified. Bergheim & Kelly (1993) described a tank with a particle trap designed by the Norwegian Hydrotechnical Laboratory. The particle trap was located within a screened section at the base of a central standpipe screen. The particle outlet flow was therefore taken from the bottom of the tank and the main outlet was near the top of the tank. Water flow through the particle outlet was 5–15% of the main flow, yet 80% of the particles were transported. The hydraulic capacity of the treatment equipment was claimed to be reduced to 2–3% of that required in tanks without oxygen addition and a separate particle outlet.

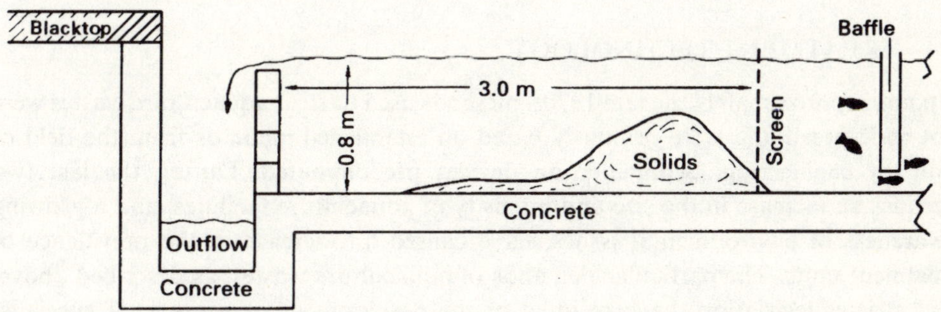

Fig. 7.7 Solid waste accumulation in the settling region of a baffled raceway (from Westers 1991).

Vertical raceways, i.e. tanks which are deeper than they are wide, can be operated so that the flow enters the tank at several levels, but flows out over a weir (Slone *et al.* (1981). This can produce a settlement region in the centre and near to the bottom of the tank. A valve located at the funnel bottom can be used to draw off the settled solids intermittently. Numerous other systems have been described, such as those by Buss *et al.* (1970), Finger (1974) and Fruchtnicht (1975). Though efficient particle collectors, not all such tank designs were likely to have been efficient at culturing animals.

The continuous removal of waste material can be beneficial, potentially producing a higher quality culture water. Such self-cleaning tanks of various shapes and sizes have been developed for many years. The principle of such tanks is that either flow dynamics are adjusted (Burrows & Chenoweth 1955, 1970), usually by the manipulation of flow rates (Cripps & Poxton 1993) or the juxtapositioning of inlet and outlet devices (Surber 1936; Burley & Klapsis 1985); that baffles are used to divert flows (Burrows & Chenoweth 1970; Wagner 1993); or that, in extreme cases, a mechanical scraper is used (Haynes 1975). While this strategy of constant removal produces a more stable effluent quality, again the resulting pollutant concentration is low and therefore difficult to treat.

In conclusion, it can be seen that traps are a useful means of collecting and concentrating waste from tanks, and that the traps must be located in a fish-free, quiescent region to aid settlement and avoid resuspension. The solids should be removed constantly, or at least frequently, to avoid leaching of nutrients, so maintaining good water quality and decreasing the proportion of the total nutrients lost in the dissolved fraction. The particles should be removed through a separate outlet specialised for that function. This will increase the concentration of solids, decrease the flow of water requiring intensive treatment, and thereby reduce the hydraulic capacity of the treatment units. A reduction in hydraulic capacity will permit the operator to choose between the purchase of a smaller and therefore cheaper device, or to increase the treatment effort. Increased effort using micro-screens could be achieved by a reduction in the screen pore size, so that more particles of a smaller size can be removed.

7.4   TREATMENT TECHNOLOGY

Up until approximately the late 1970s, methods used to treat aquaculture wastes were not widespread and were primarily based on established methods from the field of sanitary engineering. Sedimentation devices predominated. During the last two decades an increase in the size and intensity of aquaculture facilities, and a growing awareness of environmental issues, have caused an increase in the prevalence of treatment units. The particular demands of aquaculture wastes, as described above, and stricter legislation, have resulted in the development of a range of specialist treatment devices.

Within the field of waste water treatment, a specialist vocabulary has developed, but terms are commonly confused. Treatment methods which primarily utilise physical forces are unit operations, while those which are mainly achieved by chemical or biological reactions are unit processes (Tchobanoglous & Burton 1991). Unit processes and operations can be grouped into primary, secondary and tertiary treatment.

Primary treatment is the physical removal of settleable and floating solids, commonly using screening or sedimentation. This method is the most suitable for aquaculture wastes and consequently is the most common technique employed within the industry. This review will concentrate on this form of treatment. Secondary treatment refers to the biological and chemical processes used to remove organic matter. Tertiary treatment is the additional, selective removal of other constituents, such as P and N compounds, using combinations of processes and operations. Treatment systems can also be classified into seven groups (Wheaton 1977), related to the phase of the material treated: (1) liquid–liquid; (2) liquid–solid; (3) liquid–gas; (4) gas–solid; (5) gas–gas; (6) solid–solid; (7) solid–liquid–gas. Groups 1, 2, 3 and 7 are relevant to aquaculture systems.

Treatment is achieved by exploiting the differences in properties between the material to be removed from the effluent, and the bulk carrier, i.e. water. Various characteristics can be exploited (Wheaton 1977), density, particle size and electrical, chemical or magnetic properties. The first two characteristics are most commonly exploited in aquaculture systems. It is a basic principle that the treatment method employed is chosen to match the separation operation required. To accomplish this, a knowledge is required of the design and operation of equipment and the properties of the waste water.

Four main types of treatment units are commonly used within the aquaculture industry, mechanical, gravitational, chemical and biological. The first two are most common in open and flow-through systems and will be discussed in this review. Devices suitable for the treatment of aquaculture wastes have been reviewed by Wheaton (1977), Huguenin & Colt (1989), Landau (1992) and Cripps (1994).

7.4.1 Mechanical treatment

This method of treatment is becoming more widespread at the cost of sedimentation devices, as will be discussed later. A wide range of units is available, with several new models coming onto the market each year. These can be classified into three groups suitable for use in aquaculture: stationary screens, rotary screens and media filters. The media filters may also function as secondary biological filters.

Stationary screens

Screening is normally the first unit operation within a domestic treatment plant. In aquaculture facilities it may, however, be the only operation. A screen can be defined as a device with openings, generally of uniform size, that is used to retain the coarse solids found in waste water (Tchobanoglous & Burton 1991). Commonly, aqua-culture-specific screens have been adapted so that they retain fine material, with pore sizes as small as 30 μm diameter. These types of screen have recently become commonly known as microscreens. The term microscreens should theoretically refer to screens with pore sizes smaller than 1000 μm (1 mm), but in practice the term is normally reserved for pore size screens less than 300–500 μm. Microscreen units can be pressurised or unpressurised and are available in a wide range of sizes, configurations and materials (Huguenin & Colt 1989).

Stationary screens are the most basic type available. The screen is merely located across the direction of effluent flow. Coarse mesh screens, or bar racks (Fig. 7.8) in their coarsest form, can be used to stop large material damaging or blocking downstream devices, such as microscreens. Their use is though more common at farm inlets. The waste retained on the mesh or bars must be removed either manually or by the use of some form of intermittently or constantly operating cleaning device, such as a rake arm. Inclination of the screen aids the transport of solids to waste, thus maintaining the hydraulic capacity of the device, as shown by Degrémont (1973) and Tchobanoglous & Burton (1991). Stationary screens can be inexpensive and easy to construct and install, but tend to block and perforate.

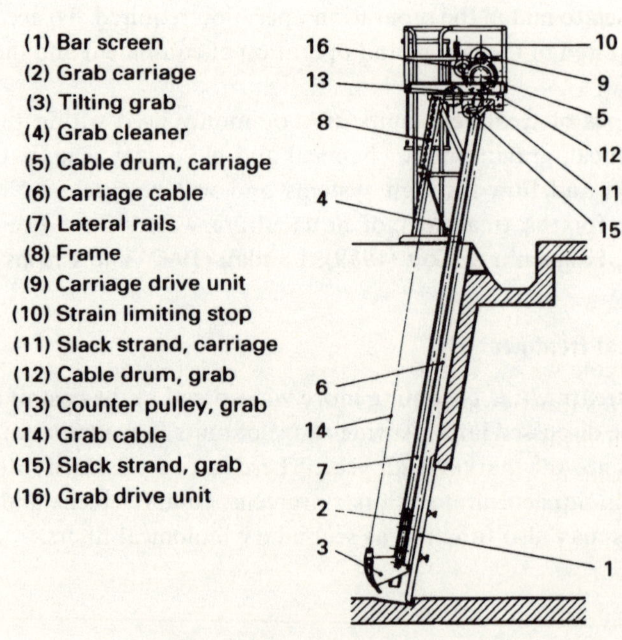

Fig. 7.8 Bar screen with rake arm (from Degrémont 1973).

A triangle filter is an advanced example of a stationary screen, suitable for aquaculture facilities, which has solved many of these problems. The farm effluent is initially passed upwards through the unit and then overflowed across the top of a screen inclined from the horizontal. The design causes the momentum of the farm effluent flow to be reduced, thus safeguarding the integrity of the screen, and encouraging the separated solids to be carried by gravity and backwash water to a waste collection trough. It is claimed by the manufacturers that 70–80% of the total phosphorus (TP) can be removed at 15–35 l/min, through a 65 μm pore size mesh. A range of pore sizes is available to suit farms with different effluent flow rates and waste particle sizes.

During tests of a 15–35 l/s capacity triangle filter (Mäkinen *et al.* 1988) fitted with a 65 μm mesh, it was estimated that the P load leaving the filter was reduced by a mean of 78%. The filter was, however, only used to treat the bottom effluent from tanks (a strategy recommended above) rather than the main flow, so the overall reduction in P loading was an average of 46%. This was stated to be a higher removal rate than that expected from a swirl concentrator.

Rotary screens

Rotary screens can be used as a substitute for primary sedimentation (Tchobanoglous & Burton 1991). They are used where blockage is likely, where it is not possible to close down an operation while the screen is cleaned, or where labour for cleaning is expensive (Wheaton 1977). In practice, in the majority of modern farms, blockage is the primary consideration, mainly because of the large flow of waste water which must pass through the screen and the small screen pore size which is required to separate out the wastes.

Rotary screens, as the name suggests, move in a circular manner as they perform their cleaning action. They are located so that part of the screen is under water, performing the separation operation, and part is above the water level, allowing this section to be backwashed or scraped, prior to entering the water again. Solids removed from the screen are directed into a waste collection trough.

The majority of rotary screens currently used in aquaculture were developed for the treatment of drinking water. The near perpendicular presentation of the screens with respect to the water flow is suited to stopping particles rather than removing them. Setting the screen at a gradient to the water flow, such as is achieved using belt screens, is a means to more gently lifting particles out of the water and directing them to waste. A further disadvantage of such devices is that the backwash sludge water flow can be substantial and may require further thickening or de-watering. The advantages are that they are easy to operate, have a low labour requirement and, in the more advanced models, the pore size, rotation speed and backwash flow can be adjusted to the particular demands of the farm. Large screens required in high flow-rate locations are difficult to rotate, and so a sequence of screens with pore sizes decreasing downstream may be required. The two main types of rotary screens are axial and radial.

The main effluent flow travels through an axial flow screen perpendicular to the plane of the screen, which is disc shaped (Fig. 7.9). If more than one screen is required, then a coarse screen is located upstream to remove larger particles and a fine screen is positioned downstream. Normally, water jets extending across the full radius of the downstream side of each screen direct cleaning water from the drinking water supply onto the screen. The waste, which accumulates on the screen during a rotation cycle, is then collected in a trough on the upstream side. The sludge streams from multiple screen units are usually combined prior to disposal or further treatment. Assuming that the hydraulic capacity of the unit and the particle removal are adequate for the application, further treatment of the main effluent flow, as opposed to the separated sludge flow, is uncommon and rarely required.

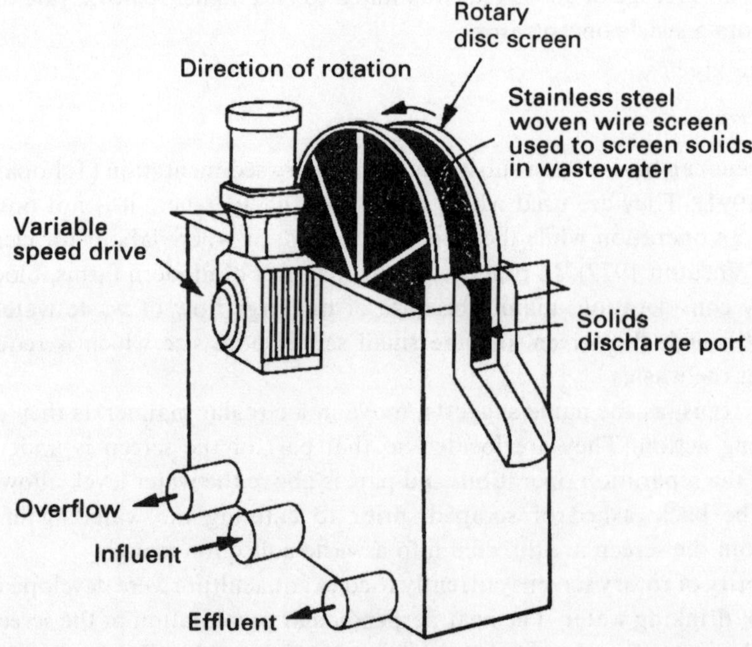

Fig. 7.9   Axial flow rotary screen (from Tchobanoglous & Burton 1991).

The *Unik Filter* (Unik Filtersystem A/S, N-5200 Os) is a commercially available example of this type of screen. Treatment efficiency estimates using this unit vary considerably, owing to variations in effluent quality and characteristics and to the pore size of the screens chosen. In a study of the waste removal efficiency using a *Unik Filter* fitted with 350 and 60 µm pore size screens, at a land-based marine salmon farm (Bergheim *et al.* 1991b), it was estimated that more than 40% of the suspended dry matter (SDM), 7–30% of the total nitrogen (TN) and more than 40% of the TP was removed. Effluent flow rates, which are typically difficult to measure, were not quantified. The low removal efficiency was attributed to the fact that only 30–40% of the TN was in the particulate form. Allowing for this, 70% of the particulate N (PN) was removed.

Liltved (1988), testing the total solids (TS), total organic carbon (TOC), TP and TN removal efficiencies using a *Unik Filter*, fitted with 1600 and 600 µm pore size screens, at two freshwater smolt farms, achieved consistently lower removal efficiencies of 20%. During tank washing operations, however, this value was increased to greater than 80%. The low removal efficiency could be attributed to the low particle concentration in the effluent and the large pore sizes of screens employed. This work confirms previous recommendations that the pre-concentration of particles prior to treatment can be a beneficial strategy.

In a comprehensive test of the *Unik Filter* (Ulgenes 1992a), removal efficiencies using a combination of 250 and 120 µm pre screens were SS (16–94%), TP (18–65%) and TN (1–49%). Efficiency was generally improved using a smaller pore size of 60 µm on the downstream screen to SS (67–73%), TP (43–74%) and TN (38–67%). These results indicate that care should be taken regarding the choice of screen pore size. Either preliminary water quality data will be required to estimate the most suitable screen needed, or the installers should offer a range of screen sizes for a commissioning period. Treatment efficiencies can then be tested *in situ* and the unsuitable screens rejected. The same *in situ* testing should also be applied to radial flow screens.

Radial flow screens are drum-shaped (Fig. 7.10). For applications with large particles, more like debris than waste feed, the effluent is passed through the screen from the outside to the inside of the drum, whilst the particles are carried over the drum to a collection channel, like a water mill (Wheaton 1977). For aquaculture applications, the effluent commonly flows first axially into the drum through one of the open ends, and then passes, radially to the axis of rotation, out through the screen

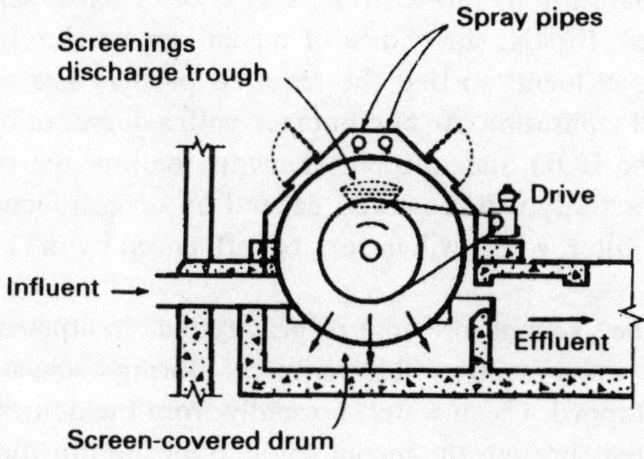

Fig. 7.10   Radial flow rotary screen (from Tchobanoglous & Burton 1991).

(Cripps 1994). The drum is located partially above the water level, so that cleaning of the drum can be facilitated. The backwash jets are located at some position out of the water on the outside of the drum, so that collected particles are washed into a trough inside the drum (Huguenin & Colt 1989). The resulting sludge is directed out of the drum through one of the open ends.

Radial flow screens have the same advantages and disadvantages as axial flow screens. It is equally important to ensure that the hydraulic capacity and pore size are matched to the characteristics of the effluent to be treated. Wheaton (1977) shows that the capacity of a radial flow screen is proportional to its length and its diameter, while the capacity of an axial flow screen is limited by the diameter. Radial flow screens are therefore not as capacity-limited as axial flow screens. In practice, however, in high flow rate applications, either a large size of unit is fitted or several units are operated in parallel. The use of parallel units has an added security advantage, because it may be possible to take one unit out of operation for repair or maintenance, while still retaining some treatment potential.

Ulgenes (1992b) tested the treatment efficiency of the Hydrotech drum filter, fitted with a 60 µm pore size screen. Treatment efficiency varied considerably within the ranges SS (67–97%), TP (21–86%) and TN (4–89%). Efficiency was found to vary proportionally with the waste effluent concentration, again indicating the advantages of particulate pre-concentration.

Media filters

Media filters, such as those containing sand, are widely used in recycle systems but are rarely used to treat the effluent from flow-through facilities, primarily because their use is most suited to lower flow situations. Their performance is strongly dependent upon the media characteristics (Huguenin & Colt 1989). Incorrect choice of media can result in only the top layers being used for particle separation. Simple designs are prone to short-circuiting and channelised flow. Provided that the hydraulic capacity of the filter is adequate for the application, this type of treatment device has two important advantages. Firstly, the choice of media can be closely matched to the characteristics of the effluent, so that the required particle size range is removed. Secondly, mechanical separation can be combined with a degree of biological activity, which will reduce the BOD and nutrient elements loading the environment. The degree of biological activity will, however, depend on several factors, including the residence time of the filter, which will in turn be influenced by the hydraulic capacity of the unit.

Media filters can be vacuum, pressure or gravity-fed. In aquaculture operations, some form of backwashing facility will be required. During backwashing the effluent flow to the filter is stopped. Clean water, normally from the domestic potable water supply, is then injected through the media in the opposite direction to the effluent flow. Trapped particles are dislodged and sent to waste. Backwashing is therefore a batch operation, so multiple parallel filters will be required if a continuous effluent treatment is required. The most advanced forms of media filter, suitable for aquaculture waste water treatment, incorporate a media washing and recycling facility (Forsell 1991).

7.4.2 Sedimentation

Sedimentation processes rely primarily upon differences in density between two media, in order for them to be separated. In aquaculture, the media are waste particles and water. Gravitational forces, in the absence of other confounding influences, cause waste matter to sink to the base of a holding vessel. In basic physical terms, the velocity is controlled by the viscosity of the media and the diameter of the particle, which is assumed to be spherical. This velocity may be predicted by Stokes' Law, although Viessman & Hammer (1993) argued that for sedimentation in urban waste water treatment systems, Stokes' Law does not hold, owing to the high level of interaction between particles. In aquaculture waste waters, where quantity may fluctuate greatly through the day, it is possible that settlement of material may at times take place in a manner predicted by Stokes' Law, however, for non-spherical particles, or where other forces such as currents and water stirring takes place, this relationship can be used only as a first estimate. Processes of flocculation, whereby waste particles either combine through natural collision and attraction, or are induced artificially, can assist settlement by increasing particle size and settling velocity. Also, through manipulation of water flow, settlement velocities may be enhanced. Settlement is one of the main methods by which the removal of suspended solids is achieved in aquaculture. The simplicity of the process and the availability of suitable sediment chambers made sedimentation one of the first forms of waste treatment in aquaculture, and remains probably the most common type of waste water treatment used at aquaculture production facilities at the moment.

Many of the first settlement devices used in aquaculture were existing earthen ponds taken out of production and used to trap effluent wastes. Settling ponds were also designed specifically to remove effluent wastes. Where ponds were designed specifically for waste treatment, retention time to allow settlement of particles is set typically at about 30 min. (Henderson & Bromage 1988) used the following relationship:

$$V_0 = \frac{Q}{A}$$

where V_o is the overflow rate (m/day), Q is the average daily flow (m^3/day) and A is the surface area of the pond (m^2).

Subsequent designs of settlement ponds sought to improve the efficiency of these systems, and have included the use of baffles to diffuse inlet water velocity and increase path length of the water flow. All such modifications are intended to have the effect of reducing water velocity within the pond, and have met with varying degrees of success.

Swirl concentrators and lamella separators are devices also considered as potential waste reduction techniques, both of which rely on settlement induced by a circulation of water within a separation chamber. Warrer-Hansen (1982) has described the design of such systems, and under certain conditions these devices have been observed to operate efficiently. The great advantage of systems using induced settlement is the need for a much-reduced installation area. Settlement ponds require considerable areas to be effective. Both lamella separators and swirl concentrators effectively reduce the required path length in which settlement can be completed. Lamella separators rely on lowering the velocity water as it is forced across plates which lie in the settlement chamber, and in effect increase the surface area of the settlement device. Laminar flow conditions are established, which permit the separation of solid and liquid phases of the waste water flow. There is a head requirement for this type of installation which may make it unsuitable for use in low-lying areas such as coastal pond systems. More importantly, the growth of fouling organisms may render this type of unit unsuitable for many aquaculture applications. A more likely application of such technology would be in systems where there is a recirculation of water.

The most pressing requirement for all sedimentation devices is the need for rapid separation of solid wastes from the water flow. While the solid particles remain in contact with the effluent stream, there exists the possibility of leaching of nutrients from the particulate to the dissolved phase where they can then be lost to the recipient. If unchecked, this release may form a significant part of the total waste load from the production system into the environment (Fig. 7.11). Aquaculture wastes are, as previously discussed, high in organic content, and as a result are susceptible to microbial and chemical processes. Excessive accumulations of waste materials in a settlement pond will lead not only to low or negative nutrient element retention, but also, through processes of gas bubble production, to resuspension of wastes and low solid matter retention. Therefore, efficiency of settlement treatment is linked intrinsically to the removal of the settled wastes at intervals from the device, which maximises removal efficiency with respect to mass of accumulated wastes and thus limits the possibilities for dissolved release.

7.4.3 Sludge

As the aquaculture industry increases in size and environmental concerns increase, waste water treatment is becoming more common. Treatment facilities for aquaculture wastes have been shown to be mainly designed to remove suspended particles, with which potentially polluting nutrients and organic matter are associated. It is therefore clear that the quantity of sludge resulting from these separation activities will increase.

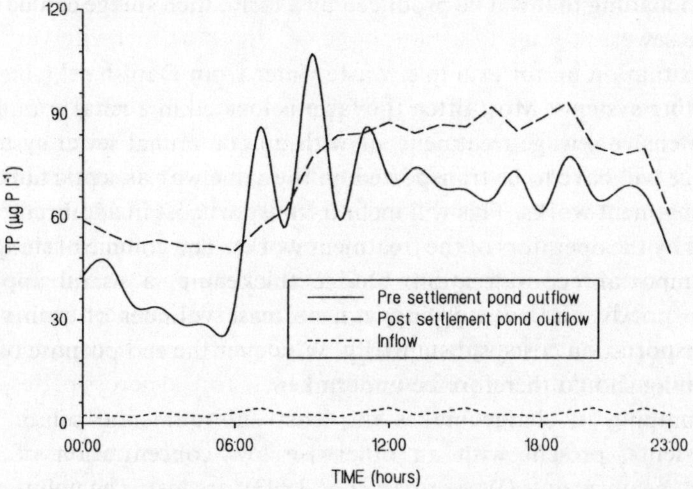

Fig. 7.11 Poorly maintained sedimentation pond generating P output through dissolution, and sustaining outflow concentration (L.A. Kelly, unpublished data).

There are two problems associated with sludge treatment. Firstly the sludge must be of an adequate density, so that minimal further thickening treatment is required. Secondly, a suitable method of disposal or re-use must be established.

It is difficult to estimate the actual quantity of sludge currently being produced, since data regarding the number of facilities employing sludge producing waste water treatment are fragmentary. Farm-specific production rates have been estimated, though many of these data concern sludge arising from the use of microscreens for the treatment of the primary effluent, rather than from other common methods such as sedimentation. The waste production from all Norwegian farms during 1990 was estimated as 8320 t N and 1440 t P (Ibrekk 1989). The actual proportion of the total number of farms employing waste water treatment techniques in general, and sludge producing techniques in particular, is currently small, primarily owing to problems associated with the collection of waste particles from under cages, but the above values do indicate the large scale of the problem. Similarly Westerman *et al.* (1993) estimated that 10 000 t of faecal wastes, which should be removed from raceways prior to discharge to the recipient water body, are produced from 23 000 t of US trout each year.

Currently, sludge treatment and disposal options available include transfer to domestic waste water treatment facilities, landfill dumping, infiltration through soil filters and use as a crop fertiliser. The first two options are the easiest to adopt, but do not make use of the rich nutrient resource in the sludge. Domestic treatment does at least break down the sludge, and so greatly reduces its impact. There is, however, usually a cost attached to this service. If the farm is located near a mains sewage system, linked to a treatment works of adequate size to cope with the loadings and fluctuations in loading that will be produced by a farm, then sludge can be discharged directly to the sewer.

This is the situation in, for example, waste water from Danish eel (*Anguilla anguilla*) recirculating systems. Most often the farm is located in a rural community with little or no intensive sewage treatment, or with no communal sewer system. In this case, the sludge will have to be transported, in the same way as septic tank liquor, by vehicle to a treatment works. This will incur a transport cost in addition to treatment charges levied by the operators of the treatment works. The volume of sludge will then become an important consideration. Sludge thickening, a useful application of sedimentation ponds, and even drying, can decrease volumes of such wastes, and therefore transportation costs, substantially. Whatever the end purpose of the waste, sludge thickening should therefore be undertaken.

Landfill dumping of sludge makes the least environmental sense. Potentially valuable nutrients, present with an otherwise low concentration of toxic contaminants, e.g. heavy metals (Westerman *et al.* 1993), are lost. The pollution problem is merely relocated. Some farms, however, own land suitable as a sludge dumping site. Here, the sludge is placed in a pit some distance from a water course. The sludge may, through leaching, eventually find its way back to the recipient water body, but, assuming the site is carefully chosen and its sludge carrying capacity not exceeded, the waste will have been filtered by the soil and biologically decomposed. This will greatly reduce, or eliminate, any aquatic impacts. This form of localised land-fill is therefore functioning as a soil filter.

The use of soil as a renovation medium for waste water has been described by Kristiansen (1982) and Westerman *et al.* (1993). Soil filters can take several forms, though three types are usually recognised, landfill dispersants, extensive crop soil filters and intensive soil filters. Land fill dispersants, as described above, are primarily a means of reducing the nutrient and BOD loading of a dumped sludge, prior to seepage back into the aquatic environment. Depending upon the suitability of the dump placement, they can also disperse the waste prior to its return to the recipient water body.

Crop soil filters are the most beneficial use to which aquaculture sludge can currently be put. The sludge is applied directly onto the land on which a plant crop is to be grown. Aquaculture sludge is particularly useful for this purpose because of the low additional contaminant concentrations, as described above, and because tests have shown (Westerman *et al.* 1993) that, for example, trout sludge is a good slow-release fertiliser, because of steady, prolonged N mineralisation. Less than 20% of the N was, however, shown to be directly available to grow crops. Cereal, vegetable or forestry crops may be suitable recipients, though in intensive coniferous woodlands care must be taken to ensure that the carrying capacity of the site is not exceeded. Stabilised sludge should not smell.

Intensive soil filters can be housed in columns, such as that described by Alanärä *et al.*, (1994), or basins. Their primary purpose is mineralisation. They are used to treat the sludge, rather than reuse or relocate it. Soil is a complex chemical, biological and physical medium. Not all naturally occurring soils are suitable for use, so columns can be filled with a specific natural or definable artificial soil with suitable properties. Whilst loading rates must be carefully limited in such intensive, relatively small volume systems, compared with crop application, soil constituents can be controlled to produce optimal mineralisation conditions.

7.5   IMPACT REDUCTION

In addition to technical solutions for the reduction in discharge quantities, planning and good farm management can aid the reduction in the quantity of waste discharged. Whilst this may therefore reduce the impact at a particular point, the overall impact of the waste may not necessarily be reduced, merely more widely dispersed. Management strategies which decrease the quantity of waste reaching a point, form non-end-of-pipe solutions to waste reduction. These solutions should be integrated with the technical treatment strategy employed to reduce mass flows, so producing the minimum possible waste water quantity and therefore maximum reduction efficiency.

At least three factors need to be considered to aid impact reduction: site selection, discharge methods together with outfall location, and sludge disposal.

In terms of site selection, surveys prior to the establishment of a site should examine the impacts of effluent composition, in addition to other environmental, economic and operational considerations. Therefore, due consideration must be given to water exchange, effluent dilution and dispersion, proximity of other sources of nutrient loading, the protection of sensitive sites (e.g. proximity to sensitive ecosystems such as coral reefs and mangroves), and multi-use conflicts (e.g. tourism and fisheries).

Discharge methods can vary and in turn produce a variable impact for the same level of waste output. A simple outfall discharging all wastes may not be adequate. Separation of particles from the main flow to allow disposal of such waste separately is the most suitable strategy in many locations. A dispersing outfall, in which the waste is passed through some form of disperser, such as a gravel substrate or perforated piping, or an outfall which discharges directly to a large recipient in the near vicinity, may, however, provide individual site solutions where surface water resources are abundant. Discharge to deep water can effectively partition the waste where it causes minimum visible damage. Care should be taken to ensure that local flow conditions do not lead this waste into unexpected distributions which may cause problems. Deep water, such as that found in fjords, can have a long residence time, so allowing the accumulation of waste products at depth, rather than safe dispersal. In the case of cage culture, the waste water flows through the sides and floor of the cage, so waste discharge strategy and reduced impact are determined more by site selection criteria than outfall design.

Sludge disposal is closely related to discharge method, as discussed above. Reuse of wastes in some form of integrated multi-crop system, or their recirculation and reconditioning in local terrestrial agricultural systems, would be a preferred method of dealing with this resource. Local conditioning and usage reduces transport and energy costs, whilst also retaining the nutrient elements within the local environment.

7.6 FUTURE TRENDS IN AQUACULTURE WASTE MANAGEMENT

This final section examines future prospects for the treatment of aquaculture waste waters. Futuristic, unorthodox and even controversial trends will be offered for discussion and criticism.

The greatest change in aquaculture waste water treatment strategy to have occurred during recent years has been merely the increase in the prevalence of some form of treatment. This change has been brought about by two main factors. The first factor is the increasing environmental awareness of the public in general, and of legislative bodies, such as environment protection agencies, in particular. The second factor concerns the increasing levels of pollution arising from a variety of sources in addition to aquaculture, and the effect this has on production and marketing.

Legislation controlling discharge permits, particularly inputs into fresh water and enclosed water bodies, has become more stringent throughout Europe. Restrictions governing a wide variety of effluent-related properties or procedures may be imposed (Table 7.4), depending on the country and region concerned. European Union (EU) policy is currently aimed at regional solutions to regional problems, rather than the more unwieldy alternative of trans-European legislation. The wide and complex

Table 7.4 Summary of the range of controls governing freshwater fish farm effluents. Adapted from Rosenthal *et al.* (1993). Prod, production capacity limitations; W treat, some form of water treatment required; N&P load, limitations on the discharge of nitrogen and phosphorus; Org load, organic loading limitations; Feed com, specified feed composition; Feed conv, limited feed conversion factor; EIS req, environmental impact statement required; Chem, regulated chemical usage; Mon, effluent monitoring required; Tax dis, tax on discharged water quantity.

	Prod	W treat	N&P load	Org load	Feed com	Feed conv	EIS req	Chem	Mon	Tax dis
Belgium			x	x						x
Denmark	x	x	x	x	x	x	x			
France			x	x			x			x
Germany		x	x	x					x	x
Greece		x	x							
Ireland	x	x	x	x			x	x	x	
Italy										
Netherlands		x		x						x
Scotland		x	x				x	x		x
Norway	x					x	x	x		
Sweden	x				x		x			
Finland	x	x	x		x	x				
US										
Canada			x	x				x		

range of laws attempting to improve what is essentially a similar problem does not, however, promote unity within the industry, or an equal platform from which to compete. In future, operational or discharge consent regulations will need to be more uniform within the EU. A recent workshop organised to discuss this subject (Rosenthal *et al.* 1993) was an indication that this problem had been identified and steps were being taken to assess options for greater uniformity.

Rosenthal *et al.* (1993) are clear that uniformity in operation procedures would be easier to achieve than uniformity in effluent characteristics. Whilst the constitution of effluents fluctuates with time and location, suitable general methods of treatment are applicable to the vast majority of similarly managed situations, though further research is urgently required to support this statement. For example, the same aperture of filter screen used in Denmark may not have the same treatment efficiency at a similar fish farm in Greece, owing to local variations in feed quality and local environmental conditions. As previously discussed, a large range of differences in management, including stock biomass and residence time, may result in differing effluent characteristics. Nevertheless, the most suitable type of treatment device may be the same in all these situations. Permitted quantities of pollutants may, conversely, be dependent on the ability of a local ecosystem to purify the compounds, or on the assessment of the importance and fragility of an ecosystem. Aquatic resource zonation (Black 1991) may be a means of tailoring treatment requirements to the local prevailing conditions.

It is therefore likely that, by the end of the 1990s, a far greater uniformity in the methods used to treat aquaculture wastes will be evident. Some form of accreditation system and a greater dissemination of independent treatment efficiency trial data from a range of devices may aid this goal. The treatment efficiency achieved from suitable devices will need not only to be adjustable to the individual environmental requirements of a given site, but also to retain the capacity for further refinement and improvement. As competition increases for the use of water resources, so aquaculture operations must be able to adapt existing systems to new demands.

Instead of effluent treatment being considered to be a negative economic activity, change in environmental protection philosophy may carry with it a change in treatment economics. In the near future it is possible that progressive aquaculture companies will be able to offset the costs of treatment against increased stock sales. Until there is a measure of treatment accreditation which conforms to the ISO 9000 systems and also national standards, the producers themselves are unlikely to realise any direct economic benefits from the process of waste water treatment. A company can use the accreditation system not only to assess the possible environmental impacts and benefits of its activities, but also to show in more general terms its economic efficiency. A company which can be shown to have few impacts, especially those resulting from waste water, or can show that it is considering such issues openly, may well be able to command a high price product or to reach markets otherwise not available (Hockin 1991; Aldridge 1994). Waste water treatment may then become a profitable occupation, further increasing the demands on current technology.

It is likely that in the near future, as indicated above, methods of treatment will become more efficient, and inefficient methods currently employed will be phased out or their function modified. One type of device which may decrease in usage within the field of aquaculture is the sedimentation pond or tank. Such passive devices can be constructed utilising unused culture facilities, and superficially appear simple to operate. Data presented here however, indicate that sedimentation tanks, when used to treat aquaculture wastes, function adequately only for brief periods in the total operational cycle.

Simple ponds, such as those used downstream of shrimp ponds, are drained and the accumulated sludge is removed intermittently. This operation is rarely undertaken more often than once at the end of each shrimp production cycle. Unless conducted extremely slowly and carefully, such a strategy will produce a large plug in nutrients, released during a short period relative to the operating period of the pond. Further, as sludge accumulates, leaching of pollutants from the sludge into the surrounding water column can occur, particularly in warm climates. Even advanced settlement pond designs with frequent (once per hour) or continuous sludge removal, are subject to hydrodynamic turbulence, resulting in the resuspension of the settled solids. Large ponds can function if residence times are sufficiently long to allow settlement of an adequate proportion of the nutrient bearing particles, and the ponds are operated with great care. More often, though, space is limiting and flow rates are high in relation to the overflow rate of the solids. Aquaculturists rarely have the time, knowledge or inclination to be sanitary engineers. In comparison, more compact units using mechanical separation techniques do not suffer from the same drawbacks.

It is more likely that, in future designs, passive sedimentation would act only as a first step in the treatment process, owing to the problems of maintaining large settlement facilities. Given the highly variable nature of waste outputs from different types of production units (Table 7.3), there would appear to be some justification for designing systems which have more than one step to the water treatment process, e.g. in-tank settlement following by screening and then sludge sedimentation. Such hybrid systems would perhaps be best suited to sites where the equipment was to be retrofitted, and work practices constrained by existing site design.

A further example indicating that care must be taken when transferring technology from the treatment of waste water from other industries or sewage is the use of flocculants. In theory, such chemicals are able to remove the small and slowly settling particles common in aquaculture effluents. Flow rates in relation to the pollutant concentration are high, and so, to achieve adequate particle separation rates, large quantities of flocculants are required. An economic assessment of flocculant application in aquaculture (Cripps 1994), showed that such use was uneconomical.

A likely consequence of increased waste water legislation is the increased use of recirculation, to a varying degree. Almost 100% recirculation, using only evaporation top-up water, is now possible even in commercial scale farms. Currently, such facilities have been limited to the production of a high-priced stock or species which require special, or stable, conditions not present in the ambient environment. Compact recycle systems commonly employ several stages of water treatment to produce water of an adequate standard for reuse. The technology suitable for this purpose has been available for many years, but capital and operational costs can be high. In areas where water volume usage, either ground or surface water, is a constraining factor, the wider introduction of this form of culture system will be imperative. Environmental requirements are altering the economic balance, so that intensive water treatment and reuse becomes more attractive. The prevalence of recycle systems in aquaculture is likely to continue to increase.

More extensive recycle systems are also becoming more popular. These involve the integrated culture of several organisms, for example fish, bivalves and seaweed (Hopkins *et al*, 1993; Kwei Lin *et al*. 1993; Shpigel *et al*., 1993). Some regard such systems as likely to fulfil the goal of achieving sustainable aquaculture. High-technology, compact treatment systems merely solve the imminent problem, that of the quantity of wastes in the waste water. Unless the separated wastes are used, the problem may only be moved downstream, or to another area such as the sludge dumping site. Ecologically-integrated systems use the waste from one culture operation as the source of nutrients for another. Thus, the overall waste is reduced. Much vital work, primarily in Israel, is continuing in this field, and such facilities should, in the long term, gradually take over from technology-intensive systems, particularly in warmer climates.

The waste water from land-based systems is currently easier to collect and treat than from open cage-based systems. As discharge restrictions increase, farm density increases and available leaseable sites become less common, it is predicted than land-based facilities will become more common. Although technically and economically the case for enclosed cage structures is not completely satisfied, there appear to be few options in cage design which will retain sufficient quantities of waste solids to satisfy pollution control authorities. The primary method by which waste treatment for cage farms in marine sites continues to be performed is through siting in areas where high levels of waste dispersal can be achieved.

Considerable improvements in sludge treatment and disposal are required. Whilst much work has been conducted assessing the efficiency of primary treatment devices, little attention has been paid to the fate of the resultant sludge. This may require further thickening and will certainly require use or disposal. It would be preferable to use the nutrients in the sludge for some purpose. If an agricultural end use for this sludge is achieved, a priority must be to have the sludge exported from the farm in a form which is readily usable by terrestrial farming systems. In the future, as research progresses and sludge quantities increase, infiltration through soil filters and use as an agricultural fertiliser will probably become more prevalent. Sludge handling must receive more attention than it does currently. A promising line of research is that concerning the fast and efficient removal of pollutants using particle traps and the location of treatment devices close to the source of the waste production, as described above.

As farm size and the stocking density of the culture organisms increase, the capital cost of the culture systems and the stock maintained increase. For security and economic reasons, the applications available for computer-controlled and automated systems are increasing. There is considerable scope for the application of continuous effluent-monitoring systems, with feedback loops to treatment devices. An application of this technology is, for example, as the quality of defined features in the effluent, e.g. TN, TP and BOD, deteriorates after feeding, the effort used to treat the effluent can be increased temporarily. An optimal treatment efficiency can be obtained and treatment effort can be maintained until the peak in poor quality water subsides. This could be achieved by varying the rotation and flushing rate of screen separators and the start-up of, or switch over to, units containing screens with a fine mesh size. Treatment could therefore be more accurately targeted at the more intense periods of waste production.

Of more immediate interest is the development of feeds customised to the treatment process. Large, or fast-sinking, particles which do not readily leach nutrients into the water column and therefore retain most of the potential pollutants in the particulate form, would be likely to improve treatment efficiency. Such beneficial treatment properties would have to be balanced against one of the main priorities of the farmer, that of stock growth. Stability of particles in the water column can oppose the needs of digestibility. The development of such a feed would nevertheless be a significant step towards an integrated farm waste management policy.

In addition to the aim of improved growth and increased food conversion efficiency, feeding frequency, ration and period can be customised to the specifications of the treatment device. Short-term plugs of high pollutant concentration effluent are easier to treat than continuous low concentrations. The adoption of a restricted feeding period (Alanärä 1992b, 1994), can also improve treatment efficiency, further improving integrated waste water treatment.

Overall, likely future trends in waste water treatment will centre around the integration of the whole culture system to produce waste water suitable for efficient treatment, or better still, a series of integrated crops which approach sustainability. Treatment devices will become more widespread and the demand for efficiency will increase, thus phasing out currently unsuitable designs.

7.7 CONCLUSIONS

- Aquaculture waste waters are difficult to treat because flow rates are high and waste concentrations are low. Total loadings may, however, be great.
- Treatment technology is aimed at the reduction in the concentration of suspended solids, phosphorus and nitrogen compounds, producing associated reductions in organic concentration and BOD.

- The characteristics of aquaculture waste waters are highly variable with time, location and site management.
- The collection of wastes from cages is currently difficult, though new methods such as closed cage systems are being developed.
- The use of particle traps in tanks, with a separate sludge outlet, is recommended because of the reduction in the hydraulic capacity and the resulting improvements in removal efficiency.
- The nutrients in aquaculture wastes are partially associated with the particulate fraction, so treatment methods which separate particles from the effluent are common in the industry.
- Rotating axial and radial flow screens are currently the most suitable group of methods commercially available for the treatment of effluents from intensive land-based aquaculture facilities.
- The fate of sludge resulting from primary effluent treatment must be considered.
- Treatment devices will become more widespread and the demand for efficiency will increase, thus phasing out currently unsuitable designs.
- In future, waste water treatment will probably centre around the integration of the whole culture system, to produce waste water suitable for efficient treatment, or better still, a series of integrated crops, approaching sustainability.

ACKNOWLEDGEMENTS

Preparation of this article was partially sponsored by the Research Council of Norway within the Programme 'Closed production systems on land at sea'. Editorial criticism and information from Dr Asbjørn Bergheim (Rogaland Research) is gratefully acknowledged.

REFERENCES

Ackefors, H. & Enell, M. (1990) Discharge of nurtrients from Swedish fish farming to adjacent sea areas. *Ambio*, **19**, 28–35.

Ackefors, H. & Enell, M. (1994) The release of nutrients and organic matter from aquaculture systems in Nordic countries. *Journal of Applied Ichthyology* **10**,(4) 225–41.

Alabaster, J.S. (1982) Survey of fish-farm effluents in some EIFAC countries. In: *Report of the EIFAC Workshop on Fish-Farm Effluents* (ed. J.S. Alabaster), pp 5–20. *EIFAC Technical Paper*, **41**.

Alanärä, A. (1992a) The effect of time restricted demand feeding activity, growth and feed conversion in rainbow trout (*Oncorhynchus mykiss*). *Aquaculture*, **108**, 357–68.

Alanärä, A. (1992b) Demand feeding as a self-regulating feeding system for rainbow trout (*Oncorhynchus mykiss*) in nets pens. *Aquaculture*, **108**. 347–56.

Alanärä, A. (1994) Demand feeding behaviour in rainbow trout. Ph.D. thesis, Swedish University of Agricultural Sciences, Umeå. ISBN 91-576-4774-7.

Alanärä, A., Bergheim, A., Cripps, S.J., Eliassen, R. & Kristiansen, R. (1994). An integrated approach to aquaculture wastewater management. *Journal of Applied Ichthyology*, **10**(4), 389.

Aldridge, C. (1994) Making the most of 'green' credentials. *Fish Farmer*, March/April, 8.

Bergheim, A. & Kelly, L. (1993) Waste treatment technologies for cage and landbased aquaculture operations. *International Conference Aquaculture and Environment*, Santiago, Chile, 2–3 September 1993. Fundación Chile: Santiago.

Bergheim, A., Aabel, J.P. & Seymour, E.A. (1991a) Past and present approaches to aquaculture waste management in Norway net pen operations. In: *Nutritional Strategies and Aquaculture Waste* (eds C.B. Cowey & C.Y. Cho). pp. 117–36. *Proceedings of the First International Symposium on Nutritional Strategies in Management of Aquaculture Waste*. University of Guelph, Guelph, Canada.

Bergheim, A., Tyvold. T. & Seymour, E.A. (1991b) Effluent loadings and sludge removal from landbased salmon farming tanks. In: *Proceedings of Aquaculture Europe '91: Aquaculture and the Environment* (compiled by N. de Pauw & J. Joyce). *EAS Special Publication*, **14**, EAS, Bredene, Belgium.

Bergheim, A., Sanni, S., Indrevik, G. & Hølland, P. (1993) Sludge removal from salmonid tank effluent using rotating microsieves. *Aquacultural Engineering*, **12**, 97–109.

Beveridge, M.C.M., Ross, L.G.R. & Kelly, L.A. (1994) Aquaculture and biodiversity. *Ambio*, **23**(8), 497–502.

Black, E.A. (1991) Coastal resource inventories: a Pacific coast strategy for aquaculture development. In: *Aquaculture and the Environment* (eds N. De Pauw & J. Joyce), pp. 441–59. *European Aquaculture Society Special Publication*, **16**, Gent, Belgium.

Boersen, G. & Westers, H. (1986) Waste solids control in hatchery raceways. *Progressive Fish Culturist*, **48**, 151–4.

Braaten, B. (1991) Impact of pollution from aquaculture in six Nordic countries. Release of nutrients, effects, and waste water treatment. In: *Aquaculture and the Environment*. (eds N. De Pauw & J. Joyce), pp. 79–101. *European Aquaculture Society Special Publication*, **16**, Gent, Belgium.

Braaten, B. (1992) Forurensning fra Nordisk akvakultur – mengder, effekter og tiltak. En statusrapport på oppdrag fra Nordisk Vattengruppe/Nordisk Ministerråd. Norsk Institutt for Vannforskning. NIVA. Oslo (In Norwegian, English abstract).

Burley, R. & Klapsis, A. (1985) Flow distribution studies in fish rearing tanks. Part 2 – Analysis of hydraulic performance of 1 m square tanks. *Aquacultural Engineering*, **4**, 113–34.

Burrows, R.E. & Chenoweth, H.H. (1955) Evaluation of three types of rearing ponds. *US Department of the Interior Fish and Wildlife Service. Research Report*, **39**.

Burrows, R.E. & Chenoweth, H.H. (1970) The rectangular circulating rearing pond. *Progressive Fish-Culturist*, April, 67–80.

Buss, K.W., Graff, D.R. & Miller, E.R. (1970) Trout culture in vertical units. *Progressive Fish-Culturist*, **32**, 187–91.

Cho, C.Y., Hynes, J.D., Wood, K.R. & Yoshida, H.K. (1991) Quantification of fish culture wastes by biological (nutritional) and chemical (limnological) methods; the development of high nutrient dense (HND) diets. In: *Nutritional Strategies and Aquaculture Waste* (eds C.B. Cowey & C.Y. Cho), pp. 37–50. University of Guelph, Guelph, Canada.

Cholette, A. & Cloutier, L. (1959) Mixing efficiency determination for continuous flow systems. *Canadian Journal of Chemical Engineering*, **37**, 105–12.

Cripps, S.J. (1993) The application of suspended particle characterization techniques to aquaculture system. In: *Techniques for Modern Aquaculture* (ed. J. Wang), 21–23 June 1993, Spokane, Washington, USA. American Society of Agricultural Engineers: St Joseph, Michigan, USA.

Cripps, S.J. (1994) Minimising outputs: treatment. *Journal of Applied Ichthyology*, **10**(4), 284–94.

Cripps, S.J. (1995) Serial particle size fractionation and characterisation of an aquacultural effluent. *Aquaculture* **133**, 323–39.

Cripps, S.J. & Poxton, M.G. (1992) A review of the design and performance of tanks relevant to flatfish culture. *Aquaculture Engineering*, **11**, 71–91.

Cripps, S.J. & Poxton, M.G. (1993) A method for the quantification and optimisation of hydrodynamics in fish culture tanks. *Aquaculture International*, **1**, 1–17.

Crozier, W.W. (1993) Evidence of genetic interaction between escaped farmed salmon and wild Atlantic salmon *(Salmo salar* L.) in a Northern Irish river. *Aquaculture*, **113**, 19–29.

Degrémont, G. (1973) *Water Treatment Handbook*. Degrémont, Rueil-Malmaison, France.

Ekbäck, D. (1975) *Rörbok – Yttre Rörledningar*. Gustavsberg, Stockholm.

Enell, M. & Ackefors, H. (1991) Belastning av fosfor och kväve, från fiskodlinger i Norden, på omgivande havsområden. *Havbrug og Miljø, Nord 1991* (eds E. Hoffmann, R. Persson, E. Gaard & G.S. Jonsson), **10**, 83–101.

Enell, M. & Lof, J. (1983) Environmental impact of aquaculture – sedimentation and nutrient loadings from fish cage culture farming. *Vatten*, **39**, 364–74.

FAO (1992) FAO Fishery Information. Aquaculture Production 1984–90. *Fish Circular*, **815**, *Rev. 4*. FAO, Rome.

Finger, J. (1974) Vertical fish farm. US Patent 3 804 063.

Fruchtnicht, E. (1975) Fish growing tank. US Patent 3 870 018.

Forsell, B. (1991) Removal of phosphorous and nitrogen from municipal wastewater by Dynasand continuous sand filter. *Environment North Seas Conference*. 26–30 May 1991, Stavanger, Norway, Vol. 2, 131–41.

Foy, R.H. & Rosell, R. (1991a) Fractionation of phosphorus and nitrogen loadings from a Northern Ireland fish farm. *Aquaculture*, **96**, 31–42.

Foy, R.H. & Rosell, R. (1991b) Loadings of nitrogen and phosphorus from a Northern Ireland fish farm. *Aquaculture*, **96**, 17–30.

GESAMP (IMO/FAO/UNESCO/WMO/WHO/IAEA/UN/UNEP (1991) Joint group of experts on the scientific aspects of marine pollution) (1991) Reducing environmental impacts of coastal aquaculture. *Report of Studies, GESAMP*, **47**.

Gowen, R.J. (1991). Aquaculture and the environment. In: *Aquaculture and the Environment* (eds N. De Pauw & J. Joyce) p 23–48. *European Aquaculture Society Special Publication*, **16**. Gent, Belgium.

Gowen, R.J., Weston, D.P. & Ervik, A. (1991) Aquaculture and the benthic environment: a review. In: *Nutritional Strategies and Aquaculture Waste* (eds C.B. Cowey & C.Y. Cho), pp. 187–205. *Proceedings of the First International Symposium on Nutritional Strategies in Management of Aquaculture Waste*. University of Guelph, Guelph, Canada.

Haynes, R. (1975) Recirculation fish raising tank system with cleanable filter. US Patent 3 886 902.

Henderson, J.P. & Bromage, N.R. (1988) Optimising the removal of suspended solids from aquacultural effluents in settlement lakes. *Aquacultural Engineering*, **7**(5), 167–81.

Henry, J.G. & Heinke, G.W. (1989) *Environmental Science and Engineering*. Prentice-Hall International Inc., London.

Hockin, D.C. (1991) Aquaculture: the case for environmental audit. In: *Aquaculture and the Environment* (eds N. De Pauw & J. Joyce), pp. 109–18. *European Aquaculture Society Special Publication*, **16**, Gent, Belgium.

Hopkins, J.S., Hamilton, R.D., Sandifer, P.A. & Browdy, C.L. (1993) The production of bivalve mollusks in intensive shrimp ponds and their effect on shrimp production and water quality. *World Aquaculture*, **24**(2), 74–7.

Howell, D.L. & Munford, J.G. (1991) Predator control on finfish farms. In: *Aquaculture and the Environment* (eds N. De Pauw & J. Joyce), pp. 339–64. *European Aquaculture Society Special Publication*, **16**. Gent, Belgium.

Huguenin, J.E. & Colt, J. (1989) *Design and Operating Guide for Aquaculture Seawater Systems*. Elsevier, Amsterdam.

Ibrekk, H.O. (1989) Model for determining pollution loading to Norwegian coastal waters (i.e. LENKA-zones). -NIVA-notat O-88145, Oslo.

Kelly, L.A. (1992) Dissolved reactive phosphorus release from sediments beneath a freshwater cage aquaculture development in West Scotland. *Hydrobiologia*, **235/236**, 569–72.

Kelly, L.A. (1993) Biological availability of phosphorus released from aquaculture wastes determined by *Selenastrum capricornutum* bioassay. *Hydrobiologia*, **253**, 367–72.

Kelly, L.A., Bergheim, A. & Hennessy, M.M. (1994) Predicting ammonium outputs from fish farms. *Water Research*, **26**(6), 1403–5.

Ketola, H.G. (1982) Effects of phosphorus in trout diets on water pollution. *Salmonid*, July – August, 12–15.

Kristiansen, R. (1982) The soil as a renovating medium – clogging of infiltration surfaces. In: *Alternative Wastewater Treatment* (eds A.S. Eikum & R.W. Seabloom). pp. 105–120. D. Reidel Publishing Company.

Kronvang, B., Ærteberg, G., Grant, R., Kristensen, P., Hovmand, M. & Kirkegaard, J. (1993) Nationwide monitoring of nutrients and their ecological effects: state of the Danish aquatic environment. *Ambio*, **22**(4), 176–87.

Kwei Lin, C., Ruamthaveesub, P. & Watnuchsoontorn, P. (1993) Integrated culture of the green mussel (*Perna viridis*) in wastewater from an intensive shrimp pond: concept and practice. *World Aquaculture*, **24**(2), 68–73.

Landau, M. (1992) *Introduction to Aquaculture*. John Wiley & Sons Inc., New York.

Liltved, H. (1988) Utprøving av Unik Hjulfilter for rensing av vann i settefiskanlegg. *NIVA VA-rapport*, **6/88**. (In Norwegian, English abstract).

Mäkinen, T., Lindgren, S. & Eskelinen, P. (1988) Sieving as an effluent treatment method for aquaculture. *Aquacultural Engineering*, **7**, 367–77.

Muir, J.F. (1982) Economic aspects of waste treatment in fish culture. In: *Report of the EIFAC Workshop on Fish-Farm Effluents* (ed. J.S. Alabaster), pp. 123–35. EIFAC *Technical Paper*, **41**.

NCC (1990) *Fish Farming and the Scottish Freshwater Environment*. Nature Conservancy Council, Edinburgh.

Penczak, T., Galicka, W., Molinski, M., Kusto, E. & Zalewski, M. (1982) The enrichment of a mesotrophic lake by carbon, phosphorus and nitrogen from the cage aquaculture of rainbow trout (*Salmo gairdneri*). *Journal of Applied Ecology*, **19**, 371–93.

Phillips, M.J. (1985) *The Environmental Impact of Cage Culture on Scottish Freshwater Lochs*. Unpublished report to the Highlands and Islands Development Board. Institute of Aquaculture, University of Stirling.

Pillay, T.V.R. (1992) *Aquaculture and the Environment*. Fishing News Books, Oxford, UK.

Pollnac, R.B. (1992) Multiuse conflicts in aquaculture – sociocultural aspects. *World Aquaculture*, **23**(2), 16–19.

Pontius, F.W. (ed.) (1990) *Water Quality and Treatment: A Handbook of Community Water Supplies* (4th edn). McGraw-Hill, New York.

Rosenthal, H., Hoffman, R., Jörgensen, L., Krüner, G., Peters, G., Schlofeldt, H.-J. & Schomann, H. (1982) Water management in circular tanks of a commercial intensive culture unit and its effects on water quality and fish condition. ICES C.M. 1982/F:22.

Rosenthal, H., Hilge, V. & Kamstra, A. (eds.) (1993) Workshop on fish farm effluents and their control in EC countries – report. Hamburg, Nov. 23–25, 1992. Christian-Albrechts-University of Kiel: Kiel, Germany. pp. 12–13.

Shepherd, C.J. & Bromage, N.R. (eds) (1988) *Intensive Fish Farming*, BSP Professional Books, Oxford.

Shpigel, M., Neori, A., Popper, D.M. & Gordin, H. (1993) A proposed model for 'environmentally clean' land-based culture of fish, bivalves and seaweeds. *Aquaculture*, **117**, 115–28.

Slone, W.J., Jester, D.B. & Turner, P.R. (1981) A closed vertical raceway fish cultural system containing cliptilolite as an ammonia stripper. In: *Proceedings of the Bio-engineering Symposium for Fish Culture* (eds. L.J. Allen & E.C. Kinney), pp. 104–15. American Fisheries Society, Bethesda, Maryland, USA.

Solbé, J.F. de L.G. (1982) Fish-farm effluents: a United Kingdom survey. In: *Report of the EIFAC Workshop on Fish-Farm Effluents* (ed. J.S. Alabaster), pp. 29–55. EIFAC *Technical paper*, **41**.

Surber, E.W. (1936) Circular rearing pools for trout and bass. *Progressive Fish-Culturist*, **21**, 1–14.

Tchobanoglous, G. & Burton, F.L. (1991) *Wastewater Engineering: Treatment, Disposal and Reuse*, 3rd Edn. McGraw-Hill Inc., New York.

Ulgenes, Y. (1992a) Undersøkelse av utslippsmengder, renseutstyr og slambehandlings-metoder ved settefiskanlegg. Delrapport III: Renseeffekt og driftserfaring med UNIK hjulfilter type 1200. *SINTEF NHL report no. STF60 A92100*. SINTEF: Trondheim, Norway.

Ulgenes, Y. (1992b) Undersøkelse av utslippsmengder, renseutstyr og slambehandlings-metoder ved settefiskanlegg. Delrapport I: Renseeffekt og driftserfaring med HYDROTECH trommelfilter. *SINTEF NHL report no. STF60 A92071*. SINTEF: Trondheim, Norway.

Ulgenes, Y. & Eikebrokk, B. (1992) Undersøkelse av utslippsmengder, renseutstyr og slambehandlingsmetoder ved settefiskanlegg. Hovedrapport. *SINTEF NHL report no. STF60 A93051*. SINTEF: Trondheim, Norway.

Viessman, W. & Hammer, M.J. (1993) *Water Supply and Pollution Control*, 5th Edn. Harper Collins College Publishers, New York.

Wagner, E.J. (1993) Evaluation of a new baffle design for solid waste removal from hatchery raceways. *Progressive Fish-Culturist*, **55**, 43–7.

Warrer-Hansen, I. (1982) Evaluation of matter discharged from trout farming in Denmark. In: *Report of the EIFAC Workshop on Fish-Farm Effluents* (ed. J.S. Alabaster), pp. 57–63. EIFAC Technical Paper, **41**.

Welch, E.B. & Lindell, T. (1992) *Ecological Effects of Wastewater: Applied Limnology and Pollutant Effects*. Chapman & Hall, London.

Westerman, P.W., Hinshaw, J.M. & Barker, J.C. (1993) Trout manure characterization and nitrogen mineralization rate. In: *Techniques for Modern Aquaculture* (ed. J. Wang), pp. 35–43. Proceedings of an Aquacultural Engineering Conference, 21–23 June 1993, Spokane, Washington, USA. American Society of Agricultural Engineers, St Joseph, Michigan, USA.

Westers, H. (1991) Operational waste management in aquaculture effluents. In: *Nutritional Strategies and Aquaculture Waste* (eds. C.B. Cowey & C.Y. Cho), pp. 231–8. *Proceedings of the First International Symposium on Nutritional Strategies in Management of Aquaculture Waste*. University of Guelph, Guelph, Canada. 275 p.

Wheaton, F.W. (1977) *Aquacultural Engineering*. John Wiley and Sons, Chichester, UK.

Wiesmann, D., Scheid, H. & Pfeffer, E. (1988) Water pollution with phosphorus of dietary origin by intensively fed rainbow trout. (*Salmo gairdneri* Rich.). *Aquaculture*, **69**, 263–70.

Chapter 8
The Legal Regime Governing Aquaculture

William R. Edeson   *Development Law Service, FAO, Rome, Italy*

8.1   INTRODUCTION

Aquaculture has been practised for many centuries, but surprisingly the legal regime governing aquaculture has only recently received detailed attention. This is quite remarkable, given that much of the aquaculture activity impinges on matters at the heart of most legal systems. It will, for example, be directly affected by the land laws, including the use of public domains such as the foreshore or mangrove areas, the water laws, environmental laws, natural resources conservation, fish and game laws, animal health and animal disease laws, as well as others applying more generally, such as public health and sanitary laws, import and export laws, tax laws, etc.

It is also interesting to contrast with this the attention received by the subject of marine fisheries, which have been a central issue in the evolution of the law of the sea over the last few centuries. This has been especially so since 1945, when attention focused on the serious issue of how the marine living resources were to be protected, as the notion that the resources of the sea were inexhaustible was being discredited by dramatic evidence of overfishing. Another important difference between the area of marine fisheries and aquaculture is that the former has been treated as a discrete area of study as a topic, whereas aquaculture by its very nature involves a consideration of the various competing uses of the resources on which aquaculture fundamentally depends, namely land and water, seed and feed, as well as the increasing awareness of its environmental dimension, and the need to ensure that aquaculture activities both respect the environment and are protected from environmental harm.

The problems of studying the legal regime of aquaculture were brought out very strongly in the first attempt to look at the subject undertaken by the Development Law Service of the FAO (Food and Agriculture Organisation of the United Nations) (Van Houtte *et al.* 1989). This study was funded by the Aquaculture Development and Coordination Programme and was based on materials obtained from the legal database at FAO headquarters, supplemented by other materials derived from selected countries. The study first of all classified countries along the following lines:

"

- those with a specific set of rules on aquaculture;
- those some specific legislation concerning aquaculture;
- those with an enabling law; and
- those with an enabling clause on aquaculture.

The study then examined the basic legal requirements for setting up an aquaculture farm.

It became apparent from this review that the area of aquaculture from a legal point of view is largely untouched, and that apart from some country-specific studies, e.g. for the UK (Howarth 1990), much work remained to be done. Recently, however, two major studies have been undertaken on the subject of the legal regime of aquaculture in Asia (Van Houtte 1995) and Europe (Van Houtte 1994).

In this paper, it is intended to look at the basic considerations from a legal point of view in formulating a legal regime for aquaculture, drawing on the studies referred to, and then examining the emerging regime in Vietnam (Edeson & Van Houtte 1994).

8.1.1   Defining aquaculture

A major threshold question regarding the study of the legal regime of aquaculture is how aquaculture should be defined. It is defined in the Oxford Dictionary as 'the cultivation of plants or breeding of animals in water', which seems straightforward enough, though for legal purposes it will be seen that such a definition is insufficient.

The Aquaculture Steering Committee of the Fisheries Department of the FAO defined aquaculture in the following terms:

'Aquaculture is the farming of aquatic organisms, including fish, molluscs, crustaceans, and aquatic plants. Farming implies some form of intervention in the rearing process to enhance production, such as regular stocking, feeding, protection from predators, etc. Farming also implies individual or corporate ownership of the stock being cultivated. For statistical purposes, aquatic organisms which are harvested by an individual or corporate body which has owned them throughout their rearing period contribute to aquaculture, while aquatic organisms which are exploitable by the public as a common property resource, with our without appropriate licences, are the harvest of fisheries.'

Here the emphasis on ownership, if only for statistical purposes, has the consequence of excluding from the definition several aquaculture activities that would almost certainly be included in the dictionary definition, or even in legal definitions, for it is the nature of the activity rather than the often rather elusive question of ownership which is the central concern.

Before a definition is prepared, it is important to determine whether the activity of aquaculture takes place on private or public land, in the sea, including brackish or fresh water, etc. Depending on where it takes place, the activity of aquaculture will be governed by a different combination of laws. For example, if it takes place entirely on private land, it could be subjected to the land laws and the laws governing the use of water, while if it takes place in the public domain, such as the foreshore, there are often special rules governing the use of that land.

However, it has to be pointed out that a legal definition of aquaculture is much more important in common law systems, where the legislative practice is to include a definition of key terms, while in civil law systems, words such as aquaculture, mariculture, pisciculture, would be applied without the need to resort to a formal definition of such terms.

In common law systems, although each situation will need to be examined separately, it would be appropriate to include not only a definition of aquaculture, but also aquatic species, aquaculture facility, and aquaculture products. Because of the possible different rules applying on private land in contrast to the public domain, it is suggested that it is also better to focus in legislation primarily on the activity of aquaculture itself, rather than on where it takes place. In this way, the legislation would avoid making what may amount to artificial distinctions between public and private property, etc. This also achieves a more consistent regime, and is especially appropriate if the intention of regulating it is to bring the activity of aquaculture for the first time under one broad legislative regime.

Following this approach, a definition of aquaculture that could be used in common law systems is one that covers 'the culture or husbandry of flora and fauna in water, whether fresh, marine or brackish, and includes any activity or activities carried out at or in connection with an aquaculture facility'. This definition will need to be backed up by a definition of aquatic species, and one possibility is to define it along the following lines: ' "aquatic species" means any aquatic flora and fauna, including all plants and animals, fish, mollusc, crustaceans, their seed and eggs.'.

Despite the importance of these two terms in the definition of aquaculture, they would probably need to be backed up by a definition of 'aquaculture facility' and 'aquaculture product'. The last two terms will become crucial in the enforcement phase of the legislation as they will most probably be central elements of the offences provided for. There is another reason for the need to go beyond a definition of 'aquaculture' itself, which is that the very nature of the activity of aquaculture requires that collateral issues are covered in the legislation. For example, because much aquaculture takes place in areas such as mangroves and tidal areas, or along the foreshore, it will be necessary to ensure that the legislative definitions are sufficiently comprehensive to include such areas. In the definitions set out above, this is achieved by the device of concentrating on the activity itself rather than on the location, though supplemented by a generic term for the location (aquaculture facility) and a generic term for the results of the aquaculture activity (the aquaculture product). Controls over the latter can be achieved by linking the permission to undertake aquaculture to particular localities and to the introduction of environmental controls over the activity of aquaculture.

8.2   CONTROLS OVER THE AQUACULTURE ACTIVITY

Studies conducted recently in the Asian region and in Europe showed that countries where a degree of aquaculture development has taken place have built up a legal framework which, in one way or another, allows for controls on the access to aquaculture activities, and provides means of preventing or curing the problem of pollution caused and suffered by aquaculture.

The diversity and complexity of the legal frameworks may depend on the legal status of the waters used (public or privately owned) (e.g. Philippines, France), on the nature of the waters (marine or freshwater) (France, Spain, Sweden, New Zealand, Norway, Republic of Korea, Malaysia), on the legal status and nature of the land used (coastal area or inland, private or public) (Sri Lanka, Madagascar, UK) on the need for a Government to regulate aquaculture in general or a specific aquaculture activity (e.g. Malaysia, France, Hong Kong, Ecuador, Singapore) and on the different questions it is called upon to deal with (use of natural/chemical feed, wild/hatched seed, etc.).

Regardless of whether or not this justifies a specific set of rules, the most widely used technique for exercising legal and administrative control over aquaculture seems to be an authorisation system whereby a governmental entity allows a person/company to operate a fish farm. This authorisation is often provided for under the fisheries legislation (e.g. Norway, UK, Israel, Cyprus, Sweden, Switzerland, Cambodia, Malaysia, Madagascar, El Salvador) or under aquaculture-specific regulations (e.g. Ireland, Hong Kong, Singapore, Malaysia, Republic of Korea, Nepal, Ecuador). In other cases it is clearly merged into the permit governing access to water and thus provided for under water management laws (e.g. Denmark, Finland, Hungary, Germany) as far as freshwater aquaculture is concerned or into authorisations required under environment protection-related legislation.

In those countries which have established a distinct procedure, under a specific set of rules, for obtaining an authorisation to set up an aquaculture farm, the procedure allows the Government authorities to limit the access to aquaculture operations in general or to specific aquaculture systems (control over activity), to direct the aquaculture development only in certain areas (control over location), and ensure environment-friendly management of the fish farm (pollution control) (Ecuador, Hong Kong, Singapore).

Government authorities may be either a central administration usually represented by the Minister of Fisheries or a person acting on his behalf (e.g. New Zealand, Israel, Ireland), or a decentralised authority at either a regional or municipal level (e.g. France, Germany, Italy, Canada). Where different interests might be involved, no authorisation will be granted without prior approval by several other agencies (e.g.

Sri Lanka, Indonesia, Hungary) or in close consultation with other competent government agencies (e.g. Denmark, Scotland, Hawaii, Hong Kong). The review undertaken in Asia and Europe evidenced that the authorisation may be named in different ways: authorisation, licence, permit, lease and concession. All of them are documents which grant a person the right to do something, and whereas these terms are often used in distinction to each other, in the area of aquaculture they often amount to the same thing. However, the term found more commonly is a licence to culture which gives the licensee, in addition to the authorisation to carry out aquaculture, the right to occupy premises or an area (water/land) for the purposes of aquaculture (Hong Kong, France, Ecuador, Madagascar), but does not operate to confer on, or vest in, a licensee any title, interest or estate in such property.

The decision to authorise is usually based on information supplied by the applicant. This seems commonly to include at least three types of document and information: (1) economical/administrative (names, label, duration requested, production target, financial sources, purpose of intended activity); (2) geographical (design of premises and facilities, location, map); and (3) technical (resources or species it uses, production system, water cycle, water quantity and quality treatment). The amount of information required and the degree of detail varies from one country to another, but it appears that the developing countries have more limited requirements than have certain developed countries. Further, the profile of the applicants may be subject to qualifications, with the result that a range of prospective applicants is likely to be eliminated (e.g. France, Mexico, Philippines, Ecuador). These qualifications might refer to the nationality, to the professional background or to policy decisions.

It may also happen that an environmental impact assessment is required (Malaysia, Sri Lanka, Indonesia), with the result that a series of other documents have to be prepared by the applicant.

Duration, payment and renewal of authorisations are likely to enhance or to limit the development of aquaculture. Indeed, potential fish farmers will be more attracted by a regime including low costs, fiscal incentives, long duration and/or automatic renewal, etc. For instance, a short duration of the authorisation and discretionary powers of the granting authority could prove to be an obstacle to the development of the aquaculture sector. From the countries examined, it appeared that aquaculture authorisations last from a minimum of 3 years to 25 years, with some exceptions like Singapore (1 year), France (35 years), and Switzerland (normally unlimited). The cost of obtaining the authorisation could be crucial. However, the information, when available, on the subject matter is quite fragmentary, but for most countries reviewed it was difficult to get relevant figures. Authorisations are usually renewable under similar terms and conditions to those under which the first was granted. Exceptionally, in some countries (e.g. Cyprus, UK), applications for renewal are considered on their merits and conditions and terms may be altered.

It is thought that ideally there should be the capability of imposing conditions on the aquaculture activity, which should cover the following: the areas in which aquaculture may be undertaken, the structure, equipment and maintenance practices to be used, the species which may be introduced into a particular aquaculture facility, the composition of the feed, or quantity that may be used, the control of the use of pharmaceutical preparations, drugs or antibiotics, the notification of diseases, disposal of dead fish, the movement of aquatic species, the monitoring and control of water quality, insurance of aquaculture facilities, maintenance of records and their content, the disclosure of information concerning aquaculture activities, the period or periods within which conditions must be fulfilled, and changes in structure, the equipment, maintenance practices.

The provision of an authorisation procedure for setting up and operating an aquaculture farm may constitute a good basis for ensuring that the farm will be managed in an environment-friendly manner. The concern regarding the environmental impacts caused and suffered by aquaculture is increasing. In fact, sound reasons exist why aquaculture ought to be environmentally regulated in its own interest and the interest of others affected by the activity. Aquaculture cannot be isolated from the wider environment in which it functions. Among environmental concerns that countries consider affecting the development of aquaculture, emphasis is put on pollution by domestic and other industrial wastes and on pesticide pollution (point and non-point source pollution). Conversely, aquaculture is apt to have adverse impacts on the environment.

In relation to the former concern (impacts suffered), the development of aquaculture appeared to bring with it two major legal and policy issues, conflicts between competing water/land uses, and water quality and quantity. The legal framework of most countries reviewed provides means of preventing or curing the problem. However, most of them are still built around traditional zoning and land-use planning laws and regulations or around recently-enacted environment-protection-related legislation, and thus not particularly specific to aquaculture. Nevertheless, some governments contribute significantly to protecting coastal water quality upon which marine/brackish water aquaculture operations rely (Hong Kong, Ecuador, Republic of Korea, France) through the designation of protected areas for the purposes of aquaculture (Republic of Korea), the creation of buffer zones around aquaculture activities within which certain developments (e.g. industrial) may not occur (e.g. Ecuador), or government planning of coastal activities along a coastal zone management plan (Sri Lanka, Thailand, France). Other countries have enacted pollution control laws and regulations which aim at protecting the water quality in the interest of propagation and protection of fish and/or aquatic life (China, Malaysia, Bangladesh).

Interestingly, few countries have met the main concern of the aquaculturist, i.e. to be entitled to compensation for having suffered an injury to his activity due to environmental changes (e.g. Republic of Korea).

Where aquaculture is likely to cause adverse impacts, the legislation of several countries provides legal tools which are intended to prevent them from occurring.

Aquaculture development is made subject to public controls over land use, site selection, coastal zone management, wetland use, control of pesticide use, etc. They aim at avoiding competing land and water uses, water quality alterations (the problem of disposal of effluents and management of the sources of waste), and at securing a healthy and marketable product. Increasingly, the use of environmental impact procedures is intended to allow for preventive measures to be formulated before the aquaculture activity starts.

8.3   THE EMERGING LEGAL REGIME GOVERNING AQUACULTURE IN VIETNAM

Although the link between aquaculture in countries such as Scotland or in other developed economies, and the legal regime governing aquaculture in Vietnam does not leap out, it is nonetheless interesting to consider briefly the legal regime governing aquaculture in the latter. Vietnam is one of many countries with centrally-planned economies which has recently undertaken fundamental changes in its legal and institutional framework in the effort to change to a market-orientated system. This reform requires an enormous effort to adopt new concepts, enact new laws and regulations, introduce new or adapt old institutions, and change the attitude of the people involved in this process.

There is no doubt that aquaculture is important for Vietnam and will increase in importance, in terms of both volume and diversity of operations. Being an important means by which food production can be increased, employment opportunities can be created and foreign exchange can be obtained, it is likely to contribute to the development of Vietnam's fisheries sector in the near future. If so, the Government will have to study the aquaculture development and its legal implications, and to face legal issues which will need to be addressed in setting up an aquaculture regime.

It appears that the main purposes of the aquaculture industry were increasing the national food supply, generating foreign exchange earnings and employment opportunities, and finally producing value-added and marketable products. The resources or species it uses are tilapia and sea-bass, crustaceans and molluscs, and some seaweeds; particular emphasis is placed on shrimp culture.

The system or elements it uses for production are: pond systems and open water systems in coastal areas, estuaries and river mouths; cage systems in lakes and rivers; and integrated fish culture with rice crop and/or animal husbandry at an experimental stage. However, the dominant system is still that of extensive culture, consisting of tidal-fed ponds that are managed under a simple system of trapping, growing and harvesting.

The environment in which the production is conducted is on low-lying inland plains, coastal swamp lands, tidal brackish areas, lakes/reservoirs in rivers and streams, along irrigation systems, and in mangrove areas.

It is obvious that a legal framework governing aquaculture operations should cover the above circumstances and enhance the development of aquaculture as well as protect the activity. However, it would not be realistic to state that a single aquaculture law could meet them. Other existing and future laws must carefully be integrated. As a matter of fact, aquaculture takes up land, uses water, deals with animals, pollutes and may receive pollution, uses chemicals etc., and thus cannot be isolated from other laws applicable in subject areas such as water and land or environment protection. Therefore, key legal issues governing this activity relate to access and use of water and land, environmental aspects such as pollution, clearing of mangroves, and fish disease.

In the present legal system, the current and future importance of aquaculture is not fully reflected. So far, very few legal texts deal with the subject matters, and whenever they do it is rather to respond to *ad hoc* needs like the use of land for aquaculture purposes and the taxes and fees to be paid by nationals and foreigners. In other words, they appear to be provisions not dealing directly with aquaculture but provisions regarding the legal conditions surrounding the setting up of an aquaculture farm.

A clear distinction is made between aquaculture and capture fishing. Indeed, firstly, statistics clearly separate aquaculture from capture fishing. Secondly, whereas aquaculture is embodied in the Land Law, which is a general and basic text dealing with land use and land allocation, capture fisheries seem to be regulated by Ministerial Ordinances. Thirdly, this suggests that aquaculture is more assimilated to an agriculture activity. A good illustration of this is that, under the Land Law, agricultural land is identified as 'the land determined to be mainly used for agricultural production such as cultivation, animal husbandry, aquatic culture or the research and experimentation on agriculture' (Article 42). Finally, the fish farmer is subject to the agriculture tax, whereas the fisherman is subject to the natural resource tax.

In order to have a comprehensive view of relevant legislation which applies to aquaculture, the legislation surrounding the setting up of a fish farm will be analysed. Therefore, questions of use and access to land and water, and the importance of environmental laws and regulations on subject activity are considered next.

8.3.1   Access to and use of land

Vietnam is a country which has paid some attention to the use of and access to land for aquaculture purposes. Appropriate government bodies are entitled to allocate land for aquaculture purposes. Under Article 5 of the Land Law, 'The State encourages the land users to invest labour, material, capital and apply scientific and technical achievements to ... reclaim virgin land, upturn uncultivated land, encroach land upon the sea, green the waste land and bare hills and coastal sandy land in order to expand areas of land for the production of agriculture, forestry, aquatic culture and salt-making.'

The fundamental principle of Vietnamese land tenure is stated in Article 1 of the

Land Law: 'Land is the property of the whole people, uniformly managed by the State'. In other words, the land is owned by the entire people of Vietnam, and the State is designated as its manager. The government organisations, from the National Assembly to the People's Committees at different levels, have the responsibility for land management. The People's Committee at the provincial level is responsible for activities concerning surveying and classification and demarcation of land. The practical work is the responsibility of the People's Committee in the District or in the Village (Article 8, Land Law).

Under Article 1 also, any person or group in Vietnamese society including foreigners, can receive land for aquaculture purposes. However, in respect of foreigners, special rules are applicable (see Chapter 5 of the Land Law).

Six categories of land are recognised under the Land Law:

(1)   agricultural land;
(2)   forest land;
(3)   residential land in rural areas;
(4)   urban land;
(5)   land for specialised use;
(6)   unused land.

Land suitable for aquaculture is embraced by agricultural land.

Land is allocated for a limited period of time. Under the Land Law, allocation occurs for 20 years for annual crops and aquatic culture, 50 years for perennial crops, with a provision for renewal if the person to whom the allocation was made still needs it and has complied with the Law.

The new Law also states clearly that the land may be made subject to 'right to exchange, transfer, lease, inherit, mortgage the land use rights' (Article 3[2]), while in Article 5, the State is to encourage the land users to invest labour, material and capital, and apply scientific and technical achievements to, amongst other matters 'increase the value of the land in its utilisation' and 'intensify farming, increase crops, and the economic effectiveness of land utilisation'.

Since 1994, a land utilisation tax is applied, which is determined on the following basis. There are six grades of agricultural land, which are classified according to the productivity of the land as shown in Table 8.1. Even though the land is used for aquaculture, it is still subject to the same rate of tax. The tax is about 6 to 7% of total output.

The procedure for the classification of the land for this purpose is set out in the Land Tax law dated 14 July 1993. A further decree dated 25 December 1993 (No. 73) explains how the land use classification is to be made for the purpose of this tax. No copy of this was available in English.

A particular regime is provided for foreign companies under the 'Regulation on rents of land, water and sea surfaces to be applied to foreign investment firms in Vietnam' (Decision No 210A-TC/VP of 1 March 1990). Foreign companies, i.e. all

Table 8.1 Classification of agricultural land in Vietnam according to productivity.

Grade	Productivity (kg rice/ha/year)
1	550
2	460
3	370
4	280
5	180
6	50

joint ventures, enterprises with 100% foreign invested capital and parties to business cooperation contracts, hereinafter called 'foreign invested capital enterprises' are subject to the following obligations:

(1) They must be allowed by the state to lease 'land, water and sea surface'; and
(2) They must pay rents. These are calculated on 'the annual base of one unit of acreage or on the whole of acreage in accordance with the lease contract'. The yearly rent for land averages from 200 to 1000 \$US/ha, for water, river, lake and gulf from 100 to 700 \$US/ha, and for sea surface from 200 to 800 \$US/km^2, and except for the case where there are acreages under unfixed use from 2000 to 10 000 \$US/km^2 [Article 2]. These rents are reduced occasionally in special cases subject to approval of the State Committee for Cooperation and Investment. The foreign invested capital enterprise shall be exempted from paying these rents when they are considered as part of the contribution of the Vietnamese shareholder to the legal capital of the company.' [Article 9: Decree 139 HDBT, Article 80 further reads: "in the case of production sharing, the income tax and other benefits of Vietnam derived from the right to use land, water surface and seasurface and any royalty, etc. shall be added to the share of the Vietnamese part of the production."]

However, under the new Land Law, it is possible that there could be an international agreement to which Vietnam is a party, which might provide for different conditions governing the use of land by a foreigner (see Chapter V of the Land Law, especially Article 82).

The land user's tenure is not unconditional during the term of allocation. General obligations are set out in Chapter III (Utilisation System for the Land Categories), the most important of which to note regarding aquaculture is the need to comply with land use planning requirements imposed, and with environmental laws. (Articles 47 and 48).

Infringements of these obligations can lead to the recovery of land by the State. These are set out in Article 20 of the Law. However, particular reference can be made to recovery where the land is used for 12 continuous months without a permit, the land user intentionally tries not to fulfil his responsibility, or the land is otherwise not used according to the purpose for which it was allocated or was not allocated by the right authority as defined under the Law.

It is difficult to judge whether this land system policy in coastal and marine areas provides for the optimisation of various activities located there as well as resolving problems of compatibility between different uses.

8.3.2   Access to and use of water

In Vietnam, water, like land, is common property, managed by the State. Little attention has apparently been paid to the regulation of the beneficial uses of water resources (i.e. use for domestic, irrigation, industrial, agriculture, etc. purposes), to the prevention of water pollution via standards and controls on industrial effluents, and finally to the quality of the water and environmental aspects. In this regard, sewage is commonly used as fertiliser, both in agriculture and aquaculture. It is even reported that there is stiff competition in some areas for sewage by fish farmers, gardeners and rice cultivators.

The right to use water resources for aquaculture purposes appears not be regulated and is nowhere defined as one of the valid competing uses of public waters. The use of water as such is apparently free.

Further, from an analysis of the Land Law, it appears that water can be an integral part of land and be allocated as such according to the procedures established in the Land Law. This could support government controls over important zones, for instance, coastal areas or estuarine habitats as important nursery grounds. It was also reported that in some areas where hydroelectric power-generating reservoirs have been built, an informal agreement is concluded between the Water and Fishery Authorities to harmonise the various uses of the water in the general interest of the community.

Again, foreign invested capital firms are subject to the payment of a royalty for the use of water unless the amount to be paid is considered as a contribution to the legal capital of the enterprise by the Vietnamese shareholder. This suggests that a concession or lease to use water can be procured from competent administrative agencies.

National planners should devise ways to encourage the economic growth of large or small scale aquaculture while protecting the environment and considering the rights of competing users of waters. Among the more important goals for uses of rivers that generally conflict with fisheries interests are agricultural development (including irrigation), industrial development (including hydroelectricity), flood control, navigation improvement, and increasing urban water supplies.

Fishery problems contingent on the multiple uses of rivers seem not to be contemplated by the law, even though different uses can generate their own set of direct and indirect impacts on fisheries. For instance, in the Mekong Basin, where a fairly comprehensive study of the possible effects of integrated development of the river basin on fisheries was conducted, it was estimated that in an 85 km stretch of the river alone, fishery losses would be about 2000 t of annual catch with a market value of $US 1.05 million. There are two decrees which should also be mentioned concerning aquaculture; first, the Fisheries Minister's decision concerning the promulgation of the regulation of the management and development of marine aquaculture, dated June 1990; and second, the decision of the Council of Ministers concerning the implementation of the law concerning the management of marine aquaculture products dated 12 June 1991. Both are available in Vietnamese only.

Regarding the water quality, no serious concern was expressed by the fish farmers and thus no strategy was taken regarding, for instance, the adoption of water quality standards particularly in relation to the culture of fish. The current legislation, apparently non-existent so far on this subject, reflects this lack of concern.

8.3.3 Environmental aspects

Recent changes in the law have introduced greater awareness of the environmental aspects of aquaculture activity, though these are dealt with in laws which impinge upon aquaculture rather than deal with it separately. First, the Land Law itself, in Article 47, stipulates that 'the use of the water surface of lakes, ponds, rivers, and canals must comply with the regulations of the environmental protection and cause no hindrance to the transportation', while Article 48 further stipulates that 'the user of land with water surface on the coastline for agricultural, forestry production, aquatic culture must conform to' amongst others, regulations which 'protect the ecosystem and the environment'. Further, Article 79 imposes on the land user the obligation to 'comply with the regulations on environmental protection, do no harm to legitimate interests of the neighbouring land users.' The Land Law also provides for land use planning that could impact on aquaculture activities, but it is not known at this stage whether there are any specific provisions governing aquaculture.

The new Law on Environment Protection provides the basis for a more comprehensive regulation of activities that could harm the environment, and although aquaculture is not specifically mentioned, it is obvious that many of the provisions will apply. In Article 2, for example, it is stated that 'Environment components include air, water resources, land, land-bed, sea, forests, plants and animals, ecosystems, living areas, production areas, national parks and landscapes'. Some specific prohibitions in the law would apply to aquaculture activity; for example Article 29(2), which prohibits amongst other things the discharge of animal and plant corpses, disease viruses into water resources, lakes, ponds, rivers, streams, canals and sea. The Law also provides penalties for breaches of the Law (Articles 50–53).

It is not possible at this stage to state how effectively this law is operating, or whether it is being used or is intended to be used in the regulation of aquaculture. However, it is evident that its provisions can have a considerable impact on aquaculture activity.

8.4 CONCLUSION

To return to our more general theme, what has emerged from our work at FAO is clear evidence that the regulation of aquaculture has tended to be approached in a very fragmented way in most countries, and this is independent of the stage of development, or indeed of the importance of aquaculture to the country. It has become clear that, where possible, we should work toward aquaculture-specific legislation, because it has its own needs and concerns which are not always properly addressed if the subject is treated merely as an appendage of other areas. This is also very important regarding the regime of the environmental management of aquaculture, for the activity of aquaculture raises specific issues not adequately covered in general environmental protection legislation. Occasionally, the need to review and revise legislation affecting aquaculture is obscured by the view that, until all the policies and institutions are in place, there is no need to review the legislation.

Although the situation needs to be assessed for each country, the point should be made that there are advantages in having the legislation in place so that as the policies and institutions develop, there is a legal framework set up to provide the basis for the implementation of the policies, and their enforcement, rather than having to wait for their subsequent introduction. A law which at least provides for an authorisation scheme and for the attachments of conditions to aquaculture, for coordination of policy, the definition of basic offences regarding the operation of an aquaculture facility, and the establishment of at least minimal safeguards for environmental protection regarding the aquaculture facility, and a basic power to make regulations for future needs is certainly better than no law at all. Of course, if policies, institutions and legislation can develop together, that is so much the better.

REFERENCES

Edeson, W.R. & Van Houtte, A.R. (1994) The legislative framework governing fisheries and aquaculture in Vietnam. *National Workshop on Environment and Aquaculture Development, Haiphong, Vietnam*, 17–19 May 1994, Annex IV–4, 488–500.

Howarth, W. (1990) The Law of Aquaculture. *The Law Relating to the Farming of Fish and Shellfish in Britain*. Fishing News Books, Oxford.

Van Houtte, A. (1994) Survey on legislation relating to licensing of fish farms and aquatic pollution control relevant for freshwater aquaculture in EIFAC member countries. European Inland Fisheries Advisory Commission, Session 18. Rome (Italy). 17–25 May 1994. FAO, Rome.

Van Houtte, A. (1995) Fundamental techniques of environmental law and aquaculture law. In: Regional Study and Workshop on Environmental Assessment and Management of Aquaculture Development. Bangkok (Thailand). 21–26 February 1994, Annex III–6, pp. 451–7. *Network of Aquaculture Centres in Asia-Pacific, Bangkok*. (*NACA*) *Environment and Aquaculture Development Series* (Thailand), **1**. FAO/ NACA, Bangkok.

Van Houtte, A.R., Bonucci, N. & Edeson, W.R. (1989) A preliminary review of selected legislation governing aquaculture. ADCP/REP/89, **42**. FAO, Rome.

Index